THE BOWL OF NIGHT

F. P. Dickson

THE BOWL OF NIGHT

The Physical Universe and Scientific Thought

The M.I.T. Press, Cambridge, Massachusetts

Library of Congress catalog card number: 71-78628

Printed in The Netherlands

FOREWORD

The life of a scientist has many attractive and admirable features, but one respect exists in which he feels greatly restricted. It is no relief to know that this restriction has been imposed by ourselves, and lies in the form which our publications can take. The scientific paper is a reasonably good vehicle for expounding in as brief a fashion as one can contrive some step forward that one may have taken, but there are demands which understandably enough the editor of any reputable journal will place on scientific papers: they must be concise and there must be no irrelevant matter. Of course, there can be some divisions of opinion about the nature of relevance, but the convention has become established that the historical outline, the philosophical rumination, even a discussion of how the author himself arrived in the first instance at his problem and then at his results, are all rather out of order.

While these restrictions may be essential and understandable in an age of rapid scientific advance, they nevertheless mean that when the scientist wishes to express himself in a broader and perhaps more intelligible fashion, or when he tries to communicate thoughts and ideas which may be of didactic importance, even if they do not represent anything startingly novel, then he cannot do so in the form of a scientific paper. Of course, one can write a book but this is a very major undertaking one does not go in for lightly.

As a result of these restrictions there exists in science a substantial body of knowledge which is referred to as "folk lore", something that is unpublishable but is expressed in lectures, in conversations, and so on. Of course, it is our own fault that we don't write books more regularly, but anybody who has ever undertaken this task knows how very much work is involved. It therefore gave me much pleasure to learn that Mr. Dickson had written a book on cosmology, a pleasure which was greatly enhanced by reading the manuscript. Here we have indeed the subject of cosmology, presented in a refreshingly personal style with a great deal of the historical development that alone makes so much of it understandable, and with a good deal of philosophical thought. Like few other fields of physics, cosmology lends itself to philosophical thought and is indeed incomplete and unintelligible without it. By taking the reader through history, showing the constant interplay of advances in local physics such as dynamics and relativity, advances in astronomy, in philosophical thought, and in depth of physical understanding, the author makes clear what a tremendous growth of human understanding has occurred in this field. Some physicists doubt the value of cosmology because there is so little in the subject that is hard, but if we value science not as a collection of facts (which it is not in any case) but as a stimulant to human thought, as a great adventure of the human race, then surely it is difficult to think of a better subject than cosmology. The observations have constantly supplied that spur to imagination, that vital driving force, which can so readily be absent in purely philosophical thought, yet just because philosophy is always involved and because we are all so much conditioned by what has gone on before, the subject is of such importance to the understanding of human advancement and of showing just how human a subject science indeed is.

Spanning the centuries with a light touch, yet never without that philosophical questioning which the attentive reader enjoys, Mr. Dickson indeed has performed a major service to the subject to which his book is devoted. Naturally, the views expressed are his own and I would not, myself, agree with every one of them, but I firmly believe that a book is fascinating to the extent to which the personality and the viewpoint of the author shows through, and I certainly believe that what is not controversial is almost certainly dull. Of that dullness there is fortunately very little in cosmology and nothing at all in Mr. Dickson's book.

Professor H. Bondi F.R.S.

PREFACE

This book is the substance of a course of lectures in elementary cosmology for students in their second or third year in the School of History and Philosophy of Science at the University of New South Wales. The purpose of the course is not to produce embryo cosmologists but to provide for students not engaged in one of the sciences with a formally mathematical basis the means to gain some understanding of the nature of scientific thought. I do not intend to imply that if one studies, for instance Physics, that an understanding of that somewhat indefinable concept, scientific thought, will automatically be acquired. As a fairly experienced employer in industry of young graduates I have been forced to the conclusion that all too many of them come out of our universities highly trained but lamentably uneducated. Training and education are by no means one and the same thing. To take an example: no amount of acquired virtuosity in the manipulation of symbols will guarantee you a mathematician. An electronic computer can outpace any human calculator and, in terms of volume of output, is a more economical proposition. Your mathematician is a man who can put meaning and, I think, a sort of life into his symbols. An educated man does more than rearrange knowledge; he adds to it. I can think of an educated man as an active shareholder in the fund of human knowledge, of which science is only one aspect.

As the greater number of students who attend my lectures and for whom this book is written are not pursuing advanced mathematical studies, I have deliberately kept it to a rather elementary mathematical level. Indeed I have found it necessary to go to some extra effort to remove the terror which mathematical formality seems to induce in many of them. This terror is not a reflection on the intellectual capacity of the students but on the way mathematics is taught at schools. It is not an easy task to treat cosmology in this way and still convey the essential ideas, but there is a peculiar interest in making the attempt. Inevitably one is involved in a treatment of some aspects of relativity which the mathematical purist would view with horror. In the context of my course I shall not apologise very loudly for what has been perpetrated, more particularly in Chapter 10. At least I have not indulged in parlour trick talk about looking straight ahead in spherical space and seeing the back of one's neck a long time ago. That kind of thing is not instructive nor particularly amusing. There are occasions when a parlour trick, like slitting a Moebius band, can be used in the lecture room to illustrate a property of surfaces, but the most pleasing reaction I ever had to that came from a student in my class several years ago. After the manner of students from time immemorial, he occasionally dissolved his academic problems in alcohol and, the evening after the lecture, took some Moebius bands to his favourite place of refreshment. There, his winnings out of topological wagers more than defrayed the cost of his potations. He also contrived to put in an acceptable examination paper at the end of the year.

In a book of this kind one does not hope to disclose exciting new knowledge of the subject nor to interpret the very latest observations. If the book turns out fairly readable, not flagrantly misleading and gives a reasonable coverage of the subject it will serve its purpose. If, in addition here and there, the material has been presented in a refreshing way, something will have been achieved. Of course it does not follow that a treatment acceptable to a class with the advantage of tutorial sessions after lectures, in which doubts and obscurities can be cleared up, will be equally acceptable as a book.

Something should be said about the notes and references. They are far from exhaustive and the selection given may seem strange to those knowledgeable in the subject. A large measure of compromise is necessary here. There are hundreds of references which any expert will know are relevant to the topics discussed, but there is no point in offering a student any reference on which he cannot readily lay his hands and the import of which may well go over his head. Unlike researchers, undergraduates simply do not have the time or the means to locate many of the texts which they ought, in one's opinion, to read. It is a chastening experience to learn from tutorial sessions that one's preconceived ideas about what they should read do not always coincide with references that really tell them something. Accordingly I have confined the references mostly to texts which are readily accessible to my students and which have proved reasonably suitable for them. It is for this reason that the rather inaccessible but important papers of J. P. Loys de Cheseaux and H. W. M. Olbers have been reproduced here as appendices.

In cosmology one does not deal much in the physical quantities used in laboratory science and technology; so far writers on the subject have not felt impelled to change from c.g.s. to m.k.s. units and I do not intend to become an innovator in this respect. The conversions are simple enough for those who wish to make them but personally, and no doubt from pure force of habit, I find that density expressed in kilogrammes per cubic metre sounds more like transformer oil than cosmic hydrogen.

The treatment of cosmology in this book is largely historical and moderately philosophical, as dictated by the policy of the School of which I have the honour to be associated. Within that policy I have been at liberty to follow my own inclinations, which are not invariably conditioned by that much praised but seldom adhered-to concept of scientific objectivity. From observations of the history of science I doubt if anyone ever made a major piece of science with objective detachment; those who did anything important were all, like artists, emotionally involved. For this freedom of approach I must thank Professor J. B. Thornton, the Head of the School, who is always helpful but never restrictive. I am also very grateful to the University of New South Wales for six months' use of a quiet room in which to work on the manuscript of this book. To Professor Bondi I am deeply indebted for permission to follow his admirable treatment of special relativity. Indeed but for reading Eddington and hearing Bondi I would never have been moved to study cosmology, still less to write any sort of book about it.

I must thank Philips Electrical Pty. Ltd., in whose service I have enjoyed over a quarter of a century, for permission to go lecturing, an example of enlightened outlook in an industrial concern. My wife has also gone far beyond any legitimate call of duty in coping with vast amounts of atrocious handwriting and mistyping. Finally Mr. P. C. Jansen has kindly made my diagrams presentable.

On second thoughts there is yet another acknowledgement to be made; to Omar Khayyám of Naishápúr, sometime mathematician to Malik Shah, Epicurean philosopher and poet. From Omar's Rubáiyát, done into English by Fitzgerald, came the mottoes of the chapters and the title of the book.

> "Awake! for Morning in the Bowl of Night
> Has flung the Stone that puts the Stars to Flight."

F. P. Dickson

CONTENTS

Into this universe, and why not knowing
Nor whence, like Water willy-nilly flowing.

CHAPTER 1

NATURE OF THE SUBJECT

Of all scientific studies none seems to have its origins more deeply rooted in the distant past than cosmology; it may even have been the parent of all science. From very early times men have tried to explain the nature and origin of the universe. The sun, moon and stars were obviously far away and the whole vast structure of the heavens, with its regular daily and yearly motions, called for explanations which had then to be in super-human terms. We have come a long way from the myths which were made thousands of years ago but it is only in the last three centuries or so that cosmology has been able to shake them off and, still more recently, it has become an established branch of science.

Cosmology today may be defined as the study of the structure and history of the universe in terms of its large scale features. It is now a part of physical science and no longer involves cosmogony, which attempts to explain why and how the universe came into existence. The word cosmos from which both cosmology and cosmogony are derived, is often used as a synonym for universe. Cosmos properly means the visible order of things and so characterises the content of the universe. Accordingly in cosmology we study the system as we see it and in cosmogony the causes which made the system what it is.

The universe is the totality of all things; it includes everything that exists and nothing can exist outside it. For something to exist it must have some duration, however small, and it must exist somewhere, that is, in the universe because there is nowhere else. In past times men have argued interminably about whether the universe exists in space and about what there was before the universe came into existence. For the cosmologist there is no need to go into these questions. Space is where things are; time is while things happen. From the philosophical viewpoint this outlook may be quite indefensible, but we need something like it to make a start in our study, however much we may modify it later. In the words of A.S. Eddington, this is a policy, not a creed.

Obviously we cannot undertake to study the whole universe in detail so we must confine ourselves to certain aspects. Let us, as proper in a field of science, consider only the matter and energy in the universe, what is called the material universe. Even this will be more than enough. [1] It is well known that matter and energy come in very small basic units. Microphysics, the study of small scale phenomena, is already a very highly developed and complex affair in which great difficulties have arisen because the exceedingly small does not behave in just the same way as things on the human scale of size.

Human beings are about halfway along the scale of sizes between the smallest and the largest objects, between electrons and galaxies. We can cope fairly well with distances and dimensions not too far removed from our own; a few million times up or down [2] in size does not present any serious difficulties, but we cannot come down to the size of atoms, let alone electrons, by direct measurement nor can we successfully apply our medium scale physical laws to these minute bodies. Perhaps electrons and protons should not even be called things. There is no individuality between particles of the same kind and the kinds seem to be only aspects of something unknowable.

In our scale we are more at home with the idea of continuity than with the discreteness we have to attribute to the ultimate units of matter and energy. We see so many things that appear to change, to move, in a smooth continuous way that continuity, involving infinite divisibility seems natural to us. Yet continuity is not just as simple as it seems. It does not mean change by infinitesimal steps such as we learned about by way of introduction to the calculus; it means change without any steps whatever.

When we talk of large scale features we mean a scale so vast that galaxies, of which our Milky Way is representative and which it would take a ray of light more than a hundred thousand years to traverse, are insignificant in size compared with the distances we have to consider. We can reasonably ignore small scale discontinuities and the local phenomena of the earth. Someday we shall have to reconcile the behaviour of the very small and the very large but we are in no position to do it yet. [3] Eddington made the attempt some thirty years ago, but it was too early and no one else is yet ready to try.

---

1. NORTH, J.D., *The Measure of the Universe,* Clarendon, 1965. P. 273.
2. The wavelength of green light is about half a micron, thus the millionth part of the length of a man's forearm.
3. EDDINGTON, A.S., *The Expanding Universe,* Pelican Books No. A70. Chapter IV.

Of course, in dealing with the very large, some of our medium scale ideas need revision. In particular we have to think of distance and of time in a different way. It is one thing to measure distances on the surface of the earth with measuring sticks and extend the span optically with theodolites as surveyors do. It is a refinement of technique but no change of principle to use the same optical method to find the distance of the moon and sun, even of the nearer stars, but it is quite another thing to plumb the depths of space by measurements of what is technically called luminosity distance. There can hardly be a more inferential kind of measurement and for things very far off it cannot be interpreted literally into distance as we know it here. Even when we have some measure of a great distance it does not tell us where the object is *now* but only where it was *then* — when its light started on its way to us ages ago. The question of where it is now, becomes meaningless.

Unless, like E.A. Milne, we try to find a few self-evident general principles and work down from them we must extend our locally derived laws of science to deal with the very large. How far extension of local laws can be safely carried is not certain, but it inevitably means a restriction on how far we can explore the Universe. (4) Whether it be finite or not we cannot inspect the whole universe by observation. The region which we can see, known as the observable universe, a region which grows in extent with technical advances in the means of observation, is for some cosmologists the only part to be considered. Most cosmological theories involve a "horizon" beyond which we cannot, even in principle, hope to see. However, other cosmologists, on the basis that what we can see ought to be a fair sample of the whole, are willing to extrapolate their ideas beyond the horizon and include in their considerations regions from which we are for ever excluded.

We must now say something about the method of cosmology, which is, in fact, normal scientific method with one or two special features in addition. A rather artificial distinction has sometimes been made between cosmology and the experimental sciences on the basis that in cosmology no experiment is possible and that it has to rely on passive observation. One could of course say the same thing about astronomy which is the source of the observations used in cosmology. The point is that experiment is nothing but the mechanics of getting an observation. Much experimental equipment is very impressive, though hardly more so than big

4. BONDI, H., *Cosmology*, Cambridge University Press, 2nd Edition, 1960. P. 10.

telescopes, and refined experimental techniques call for the utmost in skill just as do astronomical observations; but in the end it is the result, the observation, that counts.

The heart of the scientific method is the continual interplay of theory and observation. Both are essential: without observation to check it theory can degenerate into speculation while nothing can be made of observation alone. Knowledge will not somehow appear from a collection of observations without theory to give them shape and significance; ([5]) we will not even know what observations should be made.

Sir William Herschel ([6]) writing "On the construction of the Heavens" had this to say: —

> "By continuing to observe the heavens with my last constructed, and since that time much improved instrument, I am now enabled to bring more confirmation to several parts that were before but weakly supported, and also to offer a few still further extended hints, such as they present themselves to my present view. But first let me mention that, if we would hope to make any progress in an investigation of this delicate nature, we ought to avoid two opposite extremes, of which I can hardly say which is the most dangerous. If we indulge a fanciful imagination and build worlds of our own, we must not wonder at our going wide from the path of truth and nature; but these will vanish like the Cartesian vortices, that soon gave way when better theories were offered. On the other hand, if we add observation to observation, without attempting to draw not only certain conclusions, but also conjectural views from them, we offend against the very end for which only observations ought to be made. I will endeavour to keep a proper medium; but if I should deviate from that, I could wish not fall into the latter error."

There is a clear enough warning, still valid today, of the dangers of unbridled speculation and uncritical observation.

For the most part theories in cosmology are formed and handled in the same way as in other branches of science. Some set of phenomena calls for an explanation, not necessarily plausible, but logically compatible with the known facts and not inconsistent with the general run of scientific knowledge though it may contradict specific features of other

5. POPPER, K.R., *Conjectures and Refutations,* Routledge and Keegan Paul, 1963. P. 7.
6. HERSCHEL, SIR W., Phil. Trans. 1783. Reprinted in Shapley and Howarth *Source Book in Astronomy* McGraw-Hill, 1929. P. 142.

theories. If the theory has merit it will enable the theorist to predict new matters for observation which, if found to be as predicted, will lend further strength to the theory. Scientific theories are not like those propositions of logic which can be declared absolutely true or false. Indeed any proposition which can be proved incontrovertibly true will be found to contain so little information that it is practically useless. That is why the only answer to Descartes' *"Cogito ergo sum"* seems to be "So what?" With all the support a scientific theory can gain from successful predictions it is still rather like a prisoner before the court, of whom the police will say no more than that they know nothing against him. It is clear that a theory can never be proved absolutely true, for there is always the possibility that some observation as yet unknown will go against it. This is like the well known difficulty of proving a negative; one has to prove that there exists no other possibility. This problem is by no means unique to science: there is usually no legal difficulty in establishing who is the mother of a child but he would be a rash man who would undertake to prove absolutely who is the father.

Scientific theories gain strength from the verification of their predictions but, however conclusively those predictions may be verified, that is not verification of the theory itself. Each prediction verified saves the theory only for the moment until it can be tested again by what observation makes of the next prediction. Thus it is merit in a theory to be able to make many predictions by which it can be tested. The term prediction ought to be taken literally; while any theory must be consistent with established facts of observation, this is a negative kind of virtue for there is little point in covering only the same ground as existing theories. There is a danger, as Milne recognised, ([7]) in forming deductive theories starting from broad general principles. Such theories must yield results consistent with established facts and laws, but, knowing the answers in advance, one may unconsciously shape the theory in the way to give these answers when its strict development might prove it false by not giving them.

There is another way in which a theory can gain merit. At any stage in the development of a major branch of science there will be found several groups of phenomena each with a theory to account for them. If a new theory can bring these together so as to become a unified theory for all

7. MILNE, E.A., *The Fundamental Concepts of Natural Philosophy*, Proc. Roy. Soc. Edin. Vol. 62. Reprinted in MUNITZ, M.K., *Theories of the Universe* Free Press 1957. P. 358.

these phenomena, its broader explanatory power will make it better than the restricted theories. Along these lines, Popper ([8]) produces the apparently paradoxical notion that a theory of great explanatory power will be one that appears inherently implausible. Such a theory is by no means easy to make and Popper's thought is really no paradox but just a different way of stating another of his arguments against optimistic empiricism: truth is by no means manifest and is hard to come by.

As no theory can ever be proved conclusively true but all are subject to being proved false, we ought to enquire about the fate of theories which have been proved wrong. Rather few have been proved totally wrong and still fewer have been completely discarded. The phlogiston theory of heat is one such discarded theory that comes to mind. A theory which has long been known to be wrong but which has never been quite discarded is the Ptolemaic theory of the solar system. While we all agree that Copernicus was right, we are good Ptolemaists every day and talk about both sunrise and moonrise with no feeling of inconsistency. Further, the rather exacting techniques of navigation and surveying and even still more exacting geodesy, can all be treated on the basis that the heavens revolve about the earth. Copernicus has nothing more to offer the practitioners of these arts than Ptolemy, for they are concerned with the stars, not with planets.

Einstein showed that Newton was wrong in certain respects, that Newton's laws of motion do not work for velocities approaching the velocity of light but we do not completely discard Newtonian mechanics. Practically only atomic and nuclear physicists can accelerate bodies to speeds which are substantial fractions of the speed of light. The error due to using Newton's laws for things as fast as rockets fired at the moon is utterly insignificant and it is minute for the orbits of the planets. It would be pointless indeed to use the more complex calculations of special relativity for the vast number of purposes where Newtonian mechanics is adequate. What is important here is that we are well aware in how far the predictions of the two theories can safely be used.

When we come to cosmological theories we cannot "use" them or their predictions in just the same way as we can with the other theories mentioned above, they are not working theories in the sense that they can be used as the basis for developments in applied science. On the basis

---

8. POPPER, K.R., Loc. Cit. P. 58.

of cosmological theories we construct conceptual models of the universe which we can compare with the observed universe to see if they give fair representations. Because even the observed universe is very complex and contains many unknown and doubtful factors, we cannot hope to produce a model which is any more than a very sketch representation covering salient features. Even these much oversimplified models are remarkable intellectual achievements and this seems to be as far as some cosmologists want to go. Some models tend to become very formal and abstract, even to the extent of being satisfied with the development of only a metrical formula for space-time not inconsistent with observation. People who do this are in a similar position to a number of theoretical physicists who are content to elaborate only mathematical formalisms and will, in extreme cases, tell an enquirer that he is not entitled to ask what connection their formalisms may have with reality. It is not a question of intellectual snobbery nor lack of ability but a strong conviction that whatever the ultimate physical reality may be, we shall never even in principle be able to know about it. Ultimately this outlook derives from our middle position in the scale of things, from the belief that we are too far from either the very large or the very small ever really to know them.

This question of epistemology, whether we can gain genuine knowledge of reality, is one that has exercised philosophers for several thousand years and is of the utmost difficulty. Kant ([9]) was convinced that Newton's mechanics gave such knowledge but he was proved wrong, leaving the problem still unsolved ([10]). While the microphysicist is at a disadvantage because he cannot directly observe his fundamental particles, the cosmologist has at least some objects in his universe which he can learn about rather directly by observing them. Poincaré's description "what we call objective reality is in the last analysis, what is common to many thinking beings and could be common to all" is quoted by Dantzig ([11]) as being as near as we can get to the intuitive idea of reality. Since observations of the sun, moon and stars are common to most of us, and could in principle be common to all of us, we have at least a naïve ground for believing that there are external realities in the universe.

In one special way cosmological theories differ from those in other branches of science where a comparative approach is possible. There

9. POPPER, K.R., Loc. Cit. P. 93 f.
10. For cosmologists as for most other scientists the existentialist answer, that nature is what it seems to be, must be highly unsatisfactory.
11. DANTZIG, T., *Number, the Language of Science* Doubleday No. A67. P. 242.

additional checks of theories can be made by comparision with the theories of similar or analogous phenomena. Thus Huygens in his development of the longitudinal wave theory of light was able to make comparisons with the theory of sound. The very uniqueness of the universe prevents the cosmologist from using the comparative method; there is simply no basis for comparison ([12]).

Because comparison enables scientists to distinguish what are particular cases and thus to arrive at more general conclusions by an extrapolative process, this method has become very important and for many scientists is practically an article of faith. As cosmologists cannot use any such method it has been neccessary to adopt certain very broad general principles which have become virtually axiomatic.

The most prominent of these is the Cosmological Principle which states that on the large scale, the universe presents the same aspect everywhere. Apart from purely local phenomena an observer in one part of the universe will see much the same as observers in other parts. While this principle is usually associated with the name of E.A. Milne, ([13]) it is used by nearly all cosmologists in one form or another. The first clear formulation is due to Nicholas of Cusa ([14]) though it was foreshadowed by Parmenides. ([15])

Cosmologists regard the principle as a most useful regulator of theories, without which there could arise an endless variety of irrefutable hypotheses explaining in an *ad hoc* fashion observations on the local scene in terms of special properties elsewhere. The cosmological principle has often been strongly attacked, for instance by Dingle, ([16]) as an unwarrantable assumption incapable of being tested and incompatible with the comparative method used elsewhere.

The cosmological principle appears to be a member of the class of irrefutable propositions discussed by Popper ([17]) so that his method of critical rationalism should apply to it. Obviously there is no possibility of testing the principle empirically by going to other regions of the

---

12. BONDI, H., Loc. Cit. P. 10.
13. NORTH, J.D., Loc. Cit. P. 156.
14. NICHOLAS OF CUSA, *Of Learned Ignorance.* Reprinted in Munitz Loc. Cit. P. 146. "The centre and the circumference are identical." However, Whitrow, (Structure and Evolution of the Universe, Hutchinson, 1961, P. 60) gives the version "a sphere with its centre everywhere and its circumference nowhere."
15. PARMENIDES, *On Nature,* (translation in de Santillana's Origins of Scientific Thought), Mentor M.Q. 336 "Remaining the same in the self same place it abides in itself."
16. DINGLE, H., *Cosmology and Science,* Scientific American Sept. 1956 P. 234.
17. POPPER, K.R., Loc. Cit. P. 193.

universe to make comparative observations, while by definition, there can be no local testing. Faced with its utter untestability empiricists feel that the very reasonableness and desirability of the principle give it an insidious and dangerous character. However, the cosmological principle is very closely related to a group of other principles which these people commonly accept; some of them can even be classed as special cases of the cosmological principle. No one suggests that hydrogen atoms are not the same in other regions as they are here, nor that the charge of the electron differs from place to place. ([18]) In special relativity the velocity of light is the same throughout the universe. ([19]) Though Mach's principle is by no means unanimously accepted, it is agreed that there is no detectable anisotropy of inertia ([20]) caused by local masses.

It is relevant to reflect that there is an assumption similar to the cosmological principle underlying a cardinal principle of experimental science: the indefinite repeatability of experiments. It is practically an article of faith that if the same conditions are set up the same kind of result will always be obtained from the same kind of experiment. This idea is generally confined to the conditions under control at the site of the experiment but there is a hidden assumption that during a time long compared with the past and future history of human science, the earth does not move into regions where the constants of nature differ appreciably from what they are here and now.

When these and other similar ideas are put together the result is remarkably like the cosmological principle, in content as well as in status. Indeed the only important difference is that the cosmological principle overtly refers to the universe at large and postulates that on the largest scale matter is uniformly distributed throughout space. As we shall see later there are some difficult questions about how big a sample of the universe one has to consider and it is not necessarily the case that the region we can survey with our present instruments is big enough. We shall also have to consider the implications of that remarkable extension of the principle which would have the large scale distribution of matter remain unchanging in time. There is no doubt that cosmology needs its cosmolo-

---

18. McCrea, W.H., *Why are all electrons alike?*, Nature, 4932 May 9, 1964. P 537.
19. However in general relativity the velocity of light depends on the distribution of matter. As most models, including Einstein's are homogeneous, this amounts to saying that the velocity is essentially uniform throughout the universe. In models where the velocity varies with epoch, for any observer the velocity also becomes a function of distance.
20. BONDI, H., Loc. Cit. P. 30.

gical principle which in many ways operates as an over-riding principle of relativity. In fact there seems to be no workable alternative. In a rudimentary form the cosmological principle appeared almost as soon as it became possible to make a distinction between cosmogony and cosmology. While age does not necessarily ensure respectability, it will be at least interesting to see how the thinkers in antiquity were led to formulate a cosmological principle which is thus one of the oldest if not the oldest of all the principles now in use.

Dreaming while the Dawn's left hand was in the sky.

CHAPTER 2

EARLY COSMOLOGICAL HISTORY

The earliest surviving accounts of the universe come from times before the invention of writing and the scribes who wrote them down were reording traditions already ancient. It is likely that these accounts date from the Neolithic Age when people were becoming settled food producers after many millenia of nomadic food gathering. Archeologists have found the texts of ancient Mesopotamian and Egyptian cosmogonies, while the oldest known Greek versions are preserved in the epics of Homer and a somewhat later account is found in Hesiod. ([1]). Though they differ in detail according to their racial origins, these ancient cosmogonies all have the same broad structure. Not only was it necessary to explain the beginning of the world but also the origin of the gods who made things as they are.

A beginning from absolutely nothing, the Void, was inconceivable. Accordingly the cosmogonists had to postulate a formless primordial material, Chaos, to fill the Void, a material almost nothing. As Eddington has said in our own times, it is impossible to distinguish between undifferentiated sameness and non-existence. Usually water was chosen as the formless base of creation. In many ways it was a good choice; water is of itself rather formless and uniform, it is essential for life and to ancient observers it must have seemed that there was more water in the world than anything else. They could visualise the solid earth floating on the boundless ocean; rivers, lakes and springs showed there was water in the earth and there was obviously water in the sky.

Out of the water, creation was supposed to take place by a process of differentiation. Divine opposites separated in the water, where together they had been nothing, cancelling each other. Once separated, these divinities had powers which they could exercise in various ways. One of

---

1. CORNFORD, F.M., *Pattern of Ionian Cosmogony*, Reprinted in Munitz Loc. Cit. PP. 21-31.
2. There are equally interesting cosmogonies developed by the Oriental, American and Oceanic peoples but we do not refer to them here because our science is derived from the Greeks who were in turn influenced by the Middle East.

these ways was a series of partial recombinations giving rise to large families of gods and goddesses to whom could be delegated the construction of the actual world in which men live. The final establishment of world order comes about through war between the older gods and the younger. It may be that this celestial strife is based on racial traditions going back to the times when the ancestors of the cosmogonists drove out or subdued the aboriginal populations of the lands they settled.

To us these accounts may seem childish and crudely anthropomorphic but we are looking back with the advantage of knowledge and experience accumulated over thousands of years. Perhaps the contribution to our store of knowledge made by the thinkers of remote antiquity is very small but at least we have come out of the nineteenth century habit of belittling the intellectual powers of these ancients. When it had never been done before, it took as much brain power to devise a bow and arrow as it now takes for a rocket. To produce any kind of coherent account of the universe was an intellectual triumph.

Along with written cosmogony the Bronze Age saw the birth of practical astronomy. This work was done by the priests, who, apart from the conduct of religious observances, had the largely secular task of calendar making. For agriculture to be successful on the scale needed to provide food for urban populations, a reliable calendar is essential. Calendar making involves astronomical observations carried out regularly on a professional basis. This early astronomy seems to have been most highly developed in Mesopotamia, where, while Greek civilisation was still in its days of small things, the Babylonians had written records of observations going back many centuries, even to Sumerian times. It seems that when the great Thales of Miletus predicted the eclipse of 585 B.C., he relied on Babylonian astronomical data and methods of computation. [3]

Mythical cosmogony served the needs of men all through the Bronze Age, but the Iron Age brought a newer, more cosmological way of thinking. Cosmogony, like many other aspects of life in the Greek world, underwent deep changes. Iron as well as providing more effective weapons of destruction gave men more control over their environment, better agriculture, more seaworthy ships, increased trade and greater wealth for the fortunate. For us the new era of thought begins in Ionia where in the cities by the sea lived merchant princes, men of talent who could be statesmen and deep thinkers on every topic. These were the fathers of philosophy of whom the first and almost legendary greatest was Thales

---

3. DE SANTILLANA, G., Loc. Cit. P. 24.

of Miletus. While much of their cosmological thought seems very speculative to us, they were rationalists in contrast with the strict empiricism of the Babylonian astronomers.

Of course Thales and his successors could not completely sweep away the cosmogonic background of myth. There were still plenty of myths about, tales for entertainment after dinner, but for some time now these had been changing in character; the central figures were no longer culture bringers but heroes of high adventure. The heroes were men rather larger than life and surrounded by the golden haze of the past but still men, for all that they could aspire to flout the gods. Thales and his friends would pay little more than lip service to the supernatural dwellers on Olympus.

Sheer emptiness, the vacuum, was still unthinkable and Thales also chose water for his primordial material. His pupil, Anaximander, wanted something less specific than water so he postulated the Apeiron which means the unbounded, an endless nothing in particular which could become anything and everything. Thales had to support his flat earth, with the sky like a bubble over it, in the universal water but whatever the pedestal, it in turn needed another and so on in endless succession. There was once a cosmogony in which the flat earth was supported on the backs of four giant elephants which in turn stood upon the back of a mounstruous tortoise. This animal had to swim with undiminished vigour during all eternity to prevent the whole structure sinking in the waters.

Anaximander's formless Apeiron did provide a support and a means of differentiation into specific substances. Just as in water or in air, a vortex is something different and remarkably long lasting even though it is made of the same stuff as its surroundings, so in the Apeiron vortices could both be things and support things. This ingenious notion will appear again two thousand years later with Descartes. While the whirling vortex lasts a while, it does not last forever and logically minded Anaximander made this world only one of many finite worlds to come into existence and vanish after a while in the boundless, enduring Apeiron. Obviously the earth at the centre of its vortex needed no material support to keep its place and the rotation of the heavenly bodies was neatly explained. Here was a fine piece of scientific thought, working on the principle of sufficient reason; there was a good theoretical basis for the stability of the earth and so no need for an infinite regress in material support. In the same sense Anaximander was bound to admit the plurality of worlds, there being no sufficient reason for the uniqueness of this particular vortex but quite sufficient reason for many vortices.

Anaximenes, the third Ionian to publish a theory of the universe, may have been less of an abstract thinker, for he felt it better to start with a base material more like something physically knowable and chose vapour, tenuous and formless but real. Unlike the media of his predecessors, air or vapour is compressible which gave Anaximenes a way to produce the diverse substances of the world by compression and rarefaction. Highly rarefied it made fire; slightly compressed, ordinary air and by greater degrees of compression, the heavy substances. Still using the idea of vortices, he supposed that the heavy substances gathered at the centre to form the earth while light fire made the stars in the rim of the vortex.

Standing rather apart from the other Ionian philosophers was Heracleitus of Ephesus. He seems to have been a rather intolerant fellow and much given to cryptic statements. His choice of fire as the primordial element is not particularly interesting in itself but he did bring in a new idea. Instead of the differentiation of opposites out of sameness, we should think of the universe as all the same stuff, fire, in endless flux. The existence of everything consists in change and what becomes static becomes nothing. It is quite consonant with such an outlook that Heracleitus should be the first to formulate a recognisable principle of relativity. "The way up and the way down are one and the same."

Looking back with the wisdom that comes after the event, it is easy to see Anaximander as the outstanding cosmologist of Ionia, but his scheme of things ran into the troubles that come to all theories that are too far ahead of the contemporary climate of thought. It left unanswered and unanswerable a basic question of cosmogony which could not be resolved by equating God and the Apeiron. (4) Why did the vortex which made our world start? Why indeed, and how, could any vortex ever start in the utterly featureless uniformity of the Apeiron? What had really happened was that, despite all their attempts to avert it, in the final analysis the prime media of the Ionians amounted to no more than empty space. This kind of question will always arise until cosmology and cosmogony, science and theology, can separate and define their respective fields of thought.

The next school of philosophy which had something to say about cosmology was founded at Elea, a Greek colony in southern Italy, and is known by the name of Pythagoras, its famous leader. Starting with a

---

4. COLLINGWOOD, G.R., *The Idea of Nature*, Galaxy Books G.B. 31, 1961, P. 40 ff.

rather strict study of mathematics, more particularly the theory of numbers, the Pythogorean school finally ended in mysticism, but in the meantime produced some very respectable cosmology during the 5th Century B.C. Where the Ionians had come to an impasse Pythagorean thought could take another track. The vortex is not the only way to make something out of virtually nothing, for geometrical figures are empty space differentiated only by form. The Pythagoreans developed a theory that the various kinds of substances are characterised only by the geometrical form of their ultimate constituents.

We must keep in mind that early Pythagorean geometry was based on number. All sorts of figures, triangles, squares, etc. could be built up from a series of integers represented by rows and columns of dots, literally arrangements of pebbles. The shape of a number thus became an outstanding feature and it seemed that all regular shapes could be reduced to numbers. When their investigations of sound led to the discovery that musical intervals could be accurately represented by the ratios of numbers, the idea of harmony could be extended to geometry and the universe at large. Out of this concept grew the music of the spheres. In Pythagorean teaching these highly abstract theories were curiously interwoven with number magic and mysticism derived from Eastern and pre-Greek sources.

In cosmology the leading Pythagorean figures are Parmenides, Philolaus and Heracleides. Between them they elaborated ideas which were either current or recurrent for a very long time; some of them are still in use today. From Parmenides came ideas of symmetry which led him to hold that the earth is spherical and the universe spherically symmetrical about it. Parmenides did not put any faith in observation; appearances he held to be deceptive and our senses very fallible guides to the truth. Change and motion are mere appearances, the ultimate reality must be changeless, timeless and spatially immobile. Parmenides did not arrive at such implausible conclusions in any arbitrary way; (5) they came out of the logical contradictions of contemporary thought. The Pythagorean number geometry involved a kind of atomism imposed by the integer numbers, yet space itself could have no such structure, it had to be continuous. This deep dichotomy is the foundation of the famous Paradoxes of Zeno. a friend and supporter of Parmenides, and Zeno's arguments were at the time unanswerable. If something can move only in steps, unit by unit of space along its track, it cannot move at all, for there is nowhere for it

---

5. DE SANTILLANA, G., Loc. Cit. Chapter 6.

to be between units; hence space must be continuous. Not only must it be continuous, it must be isotropic, having the same properties in every direction. Further it has to be homogeneous, having the same properties everywhere. This is why Parmenides had to have a cosmological principle which he formulated by saying that the centre is everywhere but the circumference is nowhere.

Philolaus, a rather obscure genius, was the first to displace the earth from the centre of the world, not on grounds of astronomical observation, but simply because it is not worthy of such a place. Fire, the subtlest of all the elements, the source of light, heat and all life, had to be at the centre. To explain the apparent motion of the heavens he made the earth move around a central fire once in twenty-four hours in an orbit small compared with those of other bodies. Just as the moon has the same face always turned to the earth so the earth revolded with what, for the Greeks, was the inhabited side always turned away from the central fire. Sunlight was a reflection of the central fire from the surface of the sun and moonlight in turn a reflection of sunlight from the moon. It was not very long before travellers could reach the scorched uninhabitable hemisphere facing the central fire, only to find it was not like that at all. Now another body was postulated, the counter-earth, which kept station in its orbit between the earth and the central fire making the latter always invisible. It is also said that this extra body fitted in very well with Pythagorean doctrine, because, counting the sphere of stars as one body, it made the total of the celestial bodies ten, the perfect number. The writings of Philolaus have not survived so our knowledge of what he intended is at best second hand. The account of Anticthon (the counter-earth) given above is the one adopted by Dreyer, following Simplicius. (6) Another account is given by Gomperz (7), with a diagram which seems to show that he, following Orr, has missed the point. According to that diagram Anticthon would come between the central fire and the sun at night and thus, since the sun would be dark, there could be no moonlight.

Finally it remained for Heracleides of Pontus to abolish Anticthon, make the sun the central fire and there we have the Copernican solar system. This meant that Heracleides had to adopt an idea previously thought impossible, that the earth should have two motions at the same

6. DREYER, J.L.E., *History of Astronomy from Thales to Kepler*, Dover S 79, 1953, P. 42 ff.
7. GOMPERZ, T., *The Development of the Pythagorean Doctrine*, Munitz, Loc. Cit. P. 37.

time, circling yearly in its orbit and turning daily upon its own axis. While the earth clearly had to be at a considerable distance from the sun, the fact that stars showed no parallax, that is no apparent shift in position when the earth is at opposite sides of its orbit, means that the sphere of the stars must be at an enormous distance. When we compare the geocentric and heliocentric models of the universe an important aspect of cosmology appears. Both models were consistent with the facts of observation then available but the preference for the heliocentric model was based on its theoretical features.

The subsequent rejection of the heliocentric model by the very influential Aristotelians was not based on astronomical observations; the grounds were purely theoretical. Though their geocentric model has come to be named after him, Ptolemy was no abstract theoretician; he was a hardworking practical astronomer whose technical skill has very seldom been surpassed.

Like all ancient writings, the works of Aristotle are very difficult reading for a modern student lacking years of intensive classical studies. There have been so many commentators on Aristotle and so many commentators on the commentators that finding what he really meant is even more difficult than trying to get to the bottom of the higher criticism of Homer's works. Looking at a translation of Aristotle's "On the Heavens" [8] after so many years that one has forgotten most of it, it is hard to avoid feeling that the logical mechanics of the argument are of a higher level than some of the subject matter. There seems to be something specious about many of the Old Master's propositions.

Seeing that heavy things fall and light things (lighter than air) rise, that the celestial bodies move forever in circles while all other motion soon stops, Aristotle developed his propositions of natural and forced motion which inevitably puts the massive earth at the centre of things. Everything has its proper place. This makes heavy objects fall naturally in a straight line towards the centre of the earth where they naturally should be. Such a motion needs no force, it just happens naturally. Force is needed to cause an unnatural motion, say of a heavy body upwards, but no such motion can continue indefinitely. Not only does everything have a proper place to which it tends; everything occupies place at all times. Where there is no place, that is outside the universe, there can be no body. Thus place, in our terms, is a region of space bounded by the

---

8. MUNITZ, Loc. Cit. P. 89.

concave inner surface of the containing body. Since there can be no
vacuum all place is occupied by something and no material body can
move without displacing something else and occupying the place where
it was. This principle has been illustrated by the flight of an arrow, which
must keep on displacing air from its place as it flies. In the attempt to
avoid an infinite regress of consequential motions, it was suggested that
the air at the point of the arrow changed place with the air beside it
which in turn filled up the place behind it where there would otherwise
have been vacuum. It follows that every motion is caused by another
and when we have to consider the movements of the celestial bodies the
problem of the ultimate motion arises.

The universe must be finite and on the principle of perfection it must
be spherical. The sphere, having the same radius in every direction is
the only figure which can turn about one of its diameters as an axis
without involving change of place. A cube, for instance, rotating about
any axis would disturb the surrounding medium, causing continual change
of place. The orbits of the heavenly bodies had to be perfect circles
which are great circles of spheres. When observation clearly showed
that the planets did not move uniformly in perfect circles, extra spheres
were brought in with their centres suitably eccentric from the centre of
the earth to explain the various observed motions without losing the
perfection of spheres and uniform motion. It was held that each sphere
was driven in rotation by the sphere next outside it until the last moving
sphere, that of the fixed stars was reached. Beyond this came the ultimate
sphere which had to cause the motion of the sphere of stars, yet remain
itself motionless. This sphere was the Unmoved Mover which, though
logically contradictory, had to be assumed as the only alternative to an
endless succession of moving movers.

In many fields of science Aristotle's work was excellent indeed and
much more firmly based on observation than the work of Plato, from
whom many of Aristotle's ideas were derived. Yet all too often Aristotle
seems to have seen only those observations which suited his purposes
and, for cosmology at least, his doctrines had a pernicious influence for
the next thousand years and more.

There now remains only one set of ideas from the ancient world to be
considered. Atomism began with Leucippus, whose works have not survived but who seems to have been greatly overshadowed by his pupil and
successor Democritus. We have only scraps from Democritus, quoted
by other thinkers and some of their comments, largely hostile. However
his Roman follower, Lucretius, has left us a complete treatise *"De Natura*

Rerum" (⁹) which embodies in poetic form the teachings of Democritus.

From observation of the porosity of solid materials, the evaporation and condensation of water and other similar phenomena Lucretius holds that matter cannot possibly be continuous. It must consist of invisibly small particles arranged in empty space which remains in the interstices of its structure. The particles may be as small as we please but not of no size at all. They must be actual infinitesimal bodies, not geometrical points. The Aristotelian problem of place is no obstacle here; there is plenty of space for things to move in and no endless succession of displacements.

For the atomists, space is where things are or may be, not just potentially but actually infinite. If you come, says Lucretius, to the supposed end of the world you can still throw a spear ahead of you and it must go somewhere: there is no limit to where things can go. Since that which is infinite can have no centre, the earth cannot be the centre of the universe, only somewhere in it. It needs no support in space either for anywhere is as good as anywhere else.

The atoms, distributed with random motions, though on the average falling throughout space, are indestructible. It is an important principle with Lucretius that nothing can be created out of absolutely nothing nor can anything be totally destroyed. There is no creation or uncreation of atoms. They have existed and will continue to exist throughout eternity. Material bodies arise from local congregations of atoms, they wear away and vanish from loss of atoms. Every material body is bombarded by flying atoms, some stick to it, others knock some of its atoms off.

In the infinite everlasting universe of Lucretius, this world of ours cannot be unique, countless similar worlds are scattered throughout space; worlds have been forming and wearing away during the infinite past and will do so during future eternity.

Some have held that the ideas of Democritus and Lucretius came long before their time; others that they were nothing but groundless speculation but, as is usual, the truth is neither here nor there. Many modern cosmologists would be more at home in the boundless universe of Lucretius than in the strictly three-dimensional finite but perpetual spheres of Aristotle. Later on we shall meet with at least some of the ideas of Lucretius in modern form.

---

9. LUCRETIUS, *The Nature of the Universe*, Munitz, Loc. Cit. **P. 41 ff.**

Alike for those who for Today prepare
And those that after Tomorrow stare.

CHAPTER 3

THE DARK AGES

When Rome succeeded Greece as the Mediterranean power, the Romans took little interest in the mathematical and scientific philosophy of their predecessors. Perhaps the search for timeless perfection was too abstract for a nation more inclined to administration and engineering. Lucretius, like his master Democritus, was far from the main stream of Greek scientific thought, as was indeed the whole Epicurean school which in other ways appealed to many Romans. It is one of the ironies of history that Archimedes, the Greek who could have built firm intellectual bridges between pure science, applied science and technology, died at the hands of the Romans in the battle for Syracuse.

While pagan Rome was merely not interested, the rise of Christianity brought aggressive opposition to science. Beginning as a religion of the poor and the oppressed, the Christian faith had a long and bitter struggle before it became the official religion of the Roman Empire. By that time the barbarians were closing in and the battle for survival had begun. It is little wonder, then, that it was the policy of the Church to condemn and destroy all thinking of pagan origin.

This policy also involved the strictly literal acceptance of what was written in the scriptures, which meant that cosmology had to be replaced by an ancient cosmogony. Although some early Christian writers, like Clemens in the first century, did show some knowlegde of Greek science and even though some sympathy with it lingered in Alexandria until the third century, by the fourth century it had been almost forgotten. Typical of the writers of those days was the bigoted Lactantius [1] who would not hear of a spherical earth. A little later we find Cyril [2] and Severianus insisting on a flat earth and on the ancient abyss of waters, which they could not know had been inherited from pagan sources in remote

---

1. He wrote a book called "On the False Wisdom of the Philosophers" about 320 A.D.
2. This was Cyril of Jerusalem (A.D. 315-386). The better known Cyril of Alexandria who burned the library and was implicated in the murder of the lady mathematican Hypatia has been described as "a character not only un-amiable but singularly deficient in all the graces of the Christian life."

antiquity. Writers of this period firmly believed in the reality of the abyss of waters. It seems that even the great Augustine had to accept it (³) though in his studies before conversion he had acquired sympathy with other ideas.

It was not until the ninth century that authoritative writers could again openly refer to the earth as a sphere. It is not unreasonable to ask whether throughtout all these centuries there were no known thinkers other than churchmen but the answer is that there were not. Apart from the fact that there were very few educated laymen, reading and writing being confined mostly to the clergy, the Church strongly discouraged independent thinking. Thus Philoponus, an Alexandrian of the sixth century who read Aristotle and wrote a commentary on him, was considered a heretic. Though Alexandria did harbour a few free thinkers there was no place for them in Europe. Once the organisation and the wealth of the Roman Empire were gone Europe was a poor place for living. Continuous fighting and recurrent pestilence greatly reduced the population and large areas were almost empty. Life was perilous and primitive for most people while even nature turned harsher. During much of the Middle Ages the climate of Europe was considerably colder than it is now, or was in Roman times. In a thoroughly miserable life the Church offered the only hope and consolation men knew.

Meanwhile in the East Islam had risen. The first followers of the Prophet were at least as narrow minded and fanatical as the early Christian fathers had been, holding that what was not written in the Koran was not true and that what was in the Koran needed no repetition elsewhere. But there was a basic difference in the circumstances; whereas Christianity rose to power in a dying empire, Islam was the force that built a new and vigorous one.

The Arabs gained control of a vast area and as great a collection of peoples as Rome had mastered. Like the Romans, they were numerically few compared with their subject races and the Caliphs had to set up an efficient administration to control the empire they had won by military prowess. Naturally the working of the administration brought back to the rulers of the empire knowledge of the civilisations and customs of many peoples and the Caliphs, militarily secure from barbarian inroads for a

---

3. DREYER, Loc. Cit. P. 213. I have not read Augustine's *De Civitati Dei* but in his *Confessions*, (Penguin Classics L 114) he speaks of the waters in Book XII where he treats them as formless matter like the Apeiron. "This matter was itself almost nothing." and "The sky above us — you made it as a firmament to divide the waters from the waters."

long time to come, could afford to take a liberal view of things. The Caliphs themselves were essentially desert Bedouin who, in the background of their simple nomadic ancestry, quickly conceived a respect for culture. The Abbasid Caliphs particularly, wished to figure as cultured monarchs of an empire in which learning and intellectual life could flourish.

It did not take the Arabs long to discover the remains of Greek science which were eagerly taken up by men who formed a tradition of scholarship outside the ranks of the religious hierarchy. Some of these men were officials whose duties took them far from home, some travelled only from curiosity about other places. Fairly soon scholars were being made welcome even from foreign parts out of Islam, so long as they had knowledge to contribute. By and large the Arab scholars were not great innovators in the physical sciences but they were keen students, tireless translators and commentators who collected and absorbed all the knowledge they could acquire.

Christendom and Islam went their separate ways until the Crusades. For these expeditions to recover the Holy Land, inspired in part by faith, in part by feudal politics and in part by adventure, Europe paid an enormous price in human lives. Ultimately the Crusades failed in their avowed objective but, through contact with Islam, opened the doors again to worldly knowledge and ultimately to freedom of thought. The cultural one-way traffic which followed took quite a few generations to filter through the dark paths of Europe, largely by way of Spain, even after the Moors had lost their political power there. By the 13th century, in the time of Thomas Aquinas, Aristotelian philosophy was being widely taught, though not without opposition from the more conservative clergy. Indeed Aquinas himself did not escape condemnation by the reactionary Bishop of Paris, Etienne Tempier. (4)

Once Aristotle had been accepted by the Church his teachings became doctrine and then dogma. As Jacobi, one of the great mathematicians of the nineteenth century expressed it: —

"But when the Church took the sciences under her wing, she demanded that the forms in which they moved be subjected to the same unconditioned faith in authority as were her own laws. And so it happened that scholasticism, far from freeing the

---

4. Etienne Tempier, in 1277, officially condemned 219 propositions of Aristotelian philosophy including a number accepted by Aquinas. See LERNER & MAHDI *Medieval Political Philosophy*, Collier - MacMillan - P. 338.

human spirit, enchained it for many centuries to come, until the very possibility of free scientific research came to be doubted."(5)

For a span of fifteen hundred years there was conflict between religion and science with unfortunate consequences for both, which arose from policy, rather than from fundamental opposition between the natures of scientific inquiry and religious belief. After the early struggle for survival against paganism, which incidentally included scientific thought, the appalling conditions of life in the Dark Ages necessarily involved the Church in a highly authoritarian attitude, exercising as it did supreme temporal as well as spiritual power. To maintain its position, the Church was forced to establish the absolute and exclusive authority of revealed knowledge. Once we recognise the basic difference between cosmogony and cosmology and the possibility of treating them in a mutually exclusive way, we can see that the conflict, however serious, was not really fundamental but arose from circumstances now happily all but vanished.

None the less, now that the lamps of knowledge had been lit again, it was too late for authority to put them all out a second time and scholasticism, which had its peak with Aquinas, finally gave way to the Renaissance.

The first statement of cosmological significance since Lucretius was made by Nicholas, Bishop of Cusa, in the 15th century. Cusanus, as he is often called, was a scholar of independent mind, holding enough temporal power that when he declined an invitation to go to Rome and defend his ideas he could be confident that the authorities would not come to fetch him. His arguments were based on a somewhat specialised line of theological thought which led to his cosmological propositions. In his book "On Learned Ignorance", (6) Cusanus argued that the universe cannot be a material system bounded by an impenetrable outer sphere, perfect in form, having a definite radius and a fixed centre. "It is not infinite yet it cannot be conceived as finite since there are no limits within which it is enclosed." Expressed in modern terminology, Cusanus said that the universe is finite but unbounded. Equally the universe cannot have a centre but for every observer, the place where he is seems to be the centre. In other words the centre is everywhere and the circumference nowhere, a proposition equivalent to our modern

---

5. There is a story, which may be apocryphal, that a professor at the Sorbonne was shown a certain structure in a dissected body and said that if Aristotle had not declared it impossible he would have been willing to believe what he saw.
6. MUNITZ, Loc. Cit. P. 146 ff.

cosmological principle. Cusanus illustrated this as well as he could by considering observers on opposite parts of the earth, where for each, his place is the centre. A similar analogy is often used today to show what is meant by the term finite but unbounded.

On a uniform spherical surface all points are equivalent and at any point an observer will be at the centre of such part of the surface as he can survey. An apparent boundary is his horizon, but this goes with him wherever he moves and there is no real boundary to be found on the surface. Though an observer need, in principle, never come to the end of travelling on the surface of a sphere, its area is a finite quantity; just so many units in terms of the radius. Like Cusanus, we take this as an analogy to finite but unbounded space but there, for us as well, the power of visualisation stops. Space involves another dimension and we have to think of finite but unbounded volume. " A sphere in a sphere without a centre or circumference anywhere, as has been said."

Cusanus also had a principle of relativity very similar to those of later thinkers: —

> "In fact it is only by reference to something fixed that we detect the movement of anything. How would a person know that a ship was in movement if, from the ship in the middle of the river, the banks were invisible to him and he was ignorant of the fact that water flows? Therein we have the reason why every man, whether he be on the earth, on the sun or another planet, always has the impression that all other things are in movement, whilst he himself is in a sort of immovable centre."

The question of relative motion exercised the minds of a number of thinkers both before and after Cusanus, from Philoponus to Copernicus. It involved the Aristotelian idea of place in serious difficulties, for all motion is observed to take place relative to something serving as an immobile reference. Aristotle had defined place as the inner surface of the first immobile containing body. In the case of a boat at anchor in a river, neither the moving water nor the moving air could be its place, the river bed qualified as the first immobile reference. Now as the universe is in itself, containing all place, the question arises as to where is its immobile place. As all the spheres are rotating the only fixed reference point is the centre. Obviously the earth, or more strictly its centre, cannot be the containing place. The proposition of Cusanus could be interpreted so as to give a paradoxical kind of answer by considering the universe to be as much inside out as it is outside out, for by turning a sphere inside out the centre becomes the circumference. Copernicus, having set the earth in motion again, could have the sphere of stars at rest to serve as the immobile reference for all motions and thus neatly solve the problem.

Although the great astronomers of the sixteenth century, Copernicus, Brahe and Kepler were primarily interested in the solar system, their work did have interesting cosmological aspects. Is was not so much what these men actually said or wrote that is important in this field as the implications consequent upon their discoveries. When Copernicus removed the earth from the centre of the universe ([7]) he took the first step towards the modern view that our part of the universe, the part which we can observe, is typical rather than unique. The next step was soon to follow when the sun could be considered, not as the centre of the universe, but as a typical star. Meanwhile all Copernicus needed to know about the stars was that, showing no parallax, they must be at an enormous distance compared with the distances of the planets.

Tycho Brahe and Kepler shattered the celestial spheres by their observations and computations. ([8]) We have to realise that in those days the spheres carrying the planets were thought of as real impenetrable bodies witht the planets embedded in them. ([9]) It was still accepted that there could be no empty space in nature, besides which it had seemed that the planets required support and mechanical connections for their orbits. The sphere of the fixed stars, however, was kept as the boundary of the universe and for Kepler it served as the immobile standard of reference for all motions.

> "The region of the fixed stars supplies the movables with a place and a base upon which the movables are, as it were, supported; and movement is understood as taking place relative to its absolute immobility."

That Kepler may have had some idea of empty space outside the universe is suggested by his statement ([10]) that the sphere of the stars keeps the heat of the sun from flowing out.

One of the most restrictive ideas inherited from Greek thought was

---

7. It is commonly supposed that Copernicus made the system concentric with the sun but he did not. Knowing that the earth is not always at the same distance from the sun, he made the centre of its orbit lie in a position near the sun thus:
   "We therefore assert that the centre of the Earth carrying the Moon's path, passes in a great orbit among the other planets in an annual revolution round the Sun; that near the Sun is the centre of the Universe."
8. MUNITZ, Loc. Cit. P. 201.
9. To account for the observed motions on the Ptolemaic systems.
10. MUNITZ, Loc. Cit. P. 199.

that of perfection, which dictated the choice of circles and spheres as perfect figures and insisted on the changelessness of the regions beyond the sphere of the moon. The super-nova star which flared up in 1572 was very carefully observed by Tycho Brahe and others to be taken as a definite proof that change was possible in the outermost regions. Apparently the similar display in 1054 was not noticed in Europe where people were too much concerned with the next world to take much interest in this, though from Chinese records it must have been at least as spectacular as the event of 1572.

Not only did Tycho show that the comet which appeared in his time passed through the spheres but also that its orbit was not perfectly circular. Circular perfection cost Kepler many years of unnecessary work and was finally abolished by his proof that the orbits of the planets are elliptical. ([11])

The next and final step in disrupting the universe of Aristotle and Ptolemy was to break out of the sphere of the fixed stars. So far nobody but navigators and a few astronomers had taken a strong interest in the stars as such. Before the invention of the telescope, stellar observation was confined to estimating brightness and measuring the positions of stars on the celestial sphere as accurately as the available instruments would permit. This kind of positional astronomy had been practised since ancient times, one of the best star catalogues from antiquity being that of Hipparchus, a Greek predecessor of Ptolemy, around 150 B.C., the other being that of Ptolemy himself, made some 300 years later.

Shortly before the first telescopes were made two speculative thinkers published new ideas about the stars. One was an Italian monk, Giordano Bruno, the other an English mathematician, Thomas Digges. Bruno was an intemperate rebel, frequently in trouble with the authorities and was finally burnt at the stake for heresy in 1600. In England Digges' work was well received and his book, an expansion of one written by his father, went through seven editions between 1576 and 1605.

Bruno ([12]) based his argument on the logical inconsistency of the

---

11. It is interesting to reflect that the circle is the special case of the ellipse in which the major and the minor axes are exactly the same, making the two foci coincide. In the case of planetary orbits, the smallest perturbing influence would soon upset such an equality, even if it could occur. In modern times a problem of this kind arises in trying to put up a synchronous satellite which remains always over the same point of the earth's surface. Such an orbit can only be approximately maintained, by regular corrections of the motion of the satellite.

12. BRUNO, G., *On the Infinite Universe and Worlds* Munitz, Loc. Cit. P. 174 ff.

Aristotelian principle of place as applied to the universe. According to Aristotle, place is defined by the interface of the contained and containing bodies, there being no empty space. For ordinary bodies no problem arises about the inner or concave surface of the containing body but when we consider the universe there is only its convex outer surface not contained by anything. Thus he was compelled to say that the universe is contained in itself. The only conclusion Bruno could reach was that the universe must be contained in infinite space which has only the property of suitability to contain matter. Instead of having to say that the universe is nowhere because all space is inside it, we can say that it is in space. As space is limitless this region we occupy is not the only region where matter can be; where there is space for it, matter will be. Hence there can be and are an infinite number of material systems like ours. Of these systems we can see only a few, represented by the stars, each of which is a sun like ours with its system of planets. The stars are not to be considered any longer as lights stuck on a finite sphere but now as suns scattered throughout infinite space.

Digges did not work from a metaphysical basis as did Bruno but followed up logical consequences of the Copernican system. As the sphere of stars does not have to revolve once in twenty-four hours about the earth, there is no longer any sufficient reason for specifying it as an actual spherical shell of finite size.

In the days before telescopes, naked eye observations of the stars produced very inaccurate estimates of their angular diameters; the brightest stars appearing to have diameters as large as 3' of arc. This was one of the points on which Brahe based his objections to the system of Copernicus, for an apparent diameter of 3' at a distance so great that no parallax could be seen, meant that the linear diameter of the stars was at least as great as the diameter of the earth's orbit, which he held to be impossible; hence the earth could not move in an orbit. Digges however, was not concerned about this and as he could let the stars be distributed in infinite space, no problems of size on a finite sphere could arise. In his book Digges presents a diagram of the system of the world ([13]) indicating an infinite distribution of stars and noting that they are bigger and brighter than the sun. As the stars display all degrees of brightness down to the limit of visibility, Digges had reasonable observational support for his theory of their distribution. There appeared to be no reason at all why the supply of stars should run out at the limit

13. MUNITZ, Loc. Cit. P. 187.

of human sight. It was well known that any light will become invisible to the eye if the distance be made great enough.

Though both can be called speculators, there is a great difference in outlook between Bruno and Digges. The former proceeds on metaphysical bases and though he cites Copernicus to support his case, the astronomer's theory is not the real heart of his argument. Bruno's main effort is directed to refuting Aristotle; but while Digges does fire a shot in passing at the followers of Aristotle, for him Copernicus has already won the battle and Digges builds his theory firmly on the Copernican theory.

Except Digges, from Cusanus to Kepler, they all fought on two fronts. One was the struggle with established authority in which the intransigent Bruno lost his life. The other, intellectually more difficult, was the fight against the insidious *a priori* theories centring round the necessity for perfection of form and the impossibility of empty space. These theories were difficult to refute by pure logic and before the necessity for perfect circles could be swept away Brahe had to make new and better observations than had ever been made before. Even so it was not until the very end that Kepler realised which side he really was on.

It took rather longer to deal with the impossibility of vacuum. First Torricelli a pupil of Galileo showed, around 1650, that the space above the mercury in a barometer is vacuous. Then a little later Otto von Geuricke with his airpump gave a decisive and spectacular demonstration that vacuum can be produced at will.

Digges was involved in neither battle. In England he stood in no danger of being charged with heresy. There were plenty of men with new ideas in every field; science and discovery of all kinds were subjects of vivid interest and debate in an atmosphere of intellectual freedom, so Digges could safely make science in the modern sense. There was the theory of Copernicus, consistent with observation and clearly better than what had gone before; it was a good base on which to build a new and more far-reaching theory. Digges' theory was fully consistent with the observations which could be made in his own life time and for many years after; well into the 18th century it was supported by the observations made with telescopes. In fact Digges' theory as such was never formally refuted, for it was absorbed into the cosmology which developed around Newton's dynamics.

Ah Love, could thou and I conspire
To grasp this sorry Scheme of things entire.

CHAPTER 4

THE AGE OF NEWTON

With the seventeenth century we come to a time when science began to develop very rapidly on both the theoretical and the experimental sides, helped by the construction of technical apparatus of a power and precision never before known. Taking advantage of the considerable advances in mathematics which had occurred before and during the Renaissance, the new era in science was characterised by a quantitative approach, using all the known resources of mathematics and, where these were insufficient, new mathematical techniques were developed.

The questions asked about natural phenomena now take a different form. It is no longer enough to ask whether something changes but rather how much it changes and how fast. Of course, instrumental aids are useful for getting answers to these questions, but the more important thing is learning what are the right questions to ask. To make progress, there must be some coherence between the questions about any particular phenomenon and links must be established between phenomena. In this way remarkable advances were made at an early stage when experimental apparatus was still very primitive. Throughout the history of science we find that refined work of great importance has been done with simple apparatus, while no amount of beautiful equipment will compensate for lack of sophisticated thought. [1]

There is a story that Galileo dropped two balls of different weight from the leaning tower of Pisa to refute the Aristotelian theory that the heavier body would fall the faster, the times of falling being inversely proportional to the weights. [2] While such an experiment can show that there is no great difference between the times of fall, it was not then a

---

1. As a small boy I saw Rutherford using a microscope to observe the tiny flashes of light from a radioactive substance and a stopwatch to find the rate of disintegration of the atoms. Today we use elaborate and expensive electronic scintillation counters which are enormously faster and more sensitive. However, very few investigators can hope to produce results of comparable importance.
2. An experiment of this kind was made by Simon Stevin about 1586. See FORBES & DYKSTERHUIS, *A History of Science and Technology,* Pelican A 498, P. 168.

good one for finding small differences or none at all. That requires
instruments capable of measuring extremely small intervals of time, which
were not available until much later. What Galileo did was to let the
balls roll down an incline, which in principle is the same as letting them
drop vertically (3) but the time taken to descend the same distance is
longer according to the shallowness of the slope. In this way the time
scale of the experiment could be so greatly extended that even the rather
crude water clock which Galileo had to use would serve well enough.

Galileo was an ardent supporter of Copernicus, whose work he de-
fended with an uncompromising vigour which brought him into serious
conflict with the authorities at Rome. It seems to have been the daily
rotation of the earth on its axis rather than the annual revolution around
the sun that was the immediate cause of the trouble. (4) Even though
there was a difficulty in conceiving that the sphere of stars could rotate
in 24 hours because of the enormous speed involved, most people preferred to accept it because they were convinced that rotation of the earth
would have disastrous consequences. Even if the earth could carry the
air around with it and so avoid a violent perpetual gale, it was thought
that a body projected vertically upwards would not fall back on the same
spot. The idea that a body could not have more than one motion at the
one time was deeply ingrained. To this Galileo had an answer in the
form of a principle of relativity similar to those of several of his predecessors, illustrated by the example of a ship. If a ship is moving steadily
upon smooth water in a straight line, an observer in a closed cabin on
the ship would have no means of knowing whether the ship was in motion
or not. Moreover, he could move around in the cabin without reference
to any motion the ship might have but relative to the shore he would
have several motions at the one time. (5)

It was also common to argue that if the earth is rotating, centrifugal
force would throw bodies off into space. Even if bodies resting on the
surface could retain their places because of friction, then projectiles
certainly ought to be thrown off. To this also Galileo had a correct

3. Stevin had already worked out the basis of this equivalence. FORBES AND
   DYKSTERHUIS, Loc. Cit. P. 166 f.
4. The immediate grounds for the charge against Galileo related to his Copernican
   rejection of the statement in the Book of Joshua that God made the sun stand
   still for a time and even go back on its course by ten degrees.
5. When it came to the question of relative celestial motions, Galileo was
   conceptually behind Digges, for although he allowed the sphere of stars to
   have some depth, he held to the old idea of a finite outer boundary. MUNITZ,
   Loc. Cit. P. 195 f.

answer, that gravity would normally be sufficient to hold them and prevent escape. Unfortunately his mathematical approach to the problem was wrong and he thought he could show that even an infinitesimal force of gravity would be sufficient to overcome the velocity of a projectile however great.

This is not the place to treat in detail of Newtonian dynamics and the famous laws of motion but, because they lay at the foundations of cosmology for the next couple of centuries and have direct cosmological implications, there are some aspects of these laws which must be considered. Certainly no physical theory has ever been more fertile in predictions verified by observation, so much so that until the time of Einstein, it seemed infallible. But however much observation may have supported the predictions, it is not possible to derive Newton's laws from observation alone nor can they in themselves be verified by observation: their status approaches that of axioms, though they are by no means self evident.

The first law of motion is a law of inertia, saying that a body in a state of rest or of uniform motion in a straight line will continue in that state for ever unless it is acted upon by some force. It seems rather obvious that something at rest will remain at rest unless it is disturbed and reasonable that motion should go on unless there is something to alter or stop it. This is what is meant by inertia, that bodies naturally resist being moved or having their motion changed and that inertia is in direct proportion to the mass of the body, that is, the amount of matter which it contains, indeed it is the measure of the mass. Inertia is a phenomenon of common experience but Newton's law, in its full expression, is quite beyond any experience we can have. In the universe which we know there is no region where we can be sure that a body can move without being acted upon by some force or other. (6)

The states of rest or of motion which we can observe are relative. We may be at rest or in motion relative to the walls of this room, which is at rest relative to the earth which is in motion relative to the sun and so on. Introducing a technical term, this room constitutes a "frame of reference" relative to which we are now at rest. Starting from one corner, we can measure distances along the two walls and the floor which will

----

6. Even the neutrino, a particle whose existence was recently predicted and still more recently verified experimentally can have its flight interrupted. The neutrino has neither rest mass nor electrical charge, just a little energy. Of all bodies it must be the least susceptible to having its motion disturbed.

enable us to determine the position of any one of us in the room. If we retain only a position on the surface of the earth where the corner of the room now is, ignoring its material structure of walls and floor, we can still use the set of measurements, or system of co-ordinates as it is called, to determine our positions or movements but we must have an identifiable point of origin for them. Frames of reference, or relative spaces as Newton called them, are all susceptible of measurement in some way or other and for the purposes of events within them they are considered to be at rest. Frames of reference in which Newton's first law is valid are known as inertial frames since the motions of bodies in them are not accelerated; no forces are acting against the inertia of the bodies.

Taking the process of nomination of frames of reference to the limit, there ought to be a final frame, a universal standard of rest to which all relative states of rest or motion could ultimately be referred. If we take the hypothetical case of two bodies in uniform relative motion, carrying observers who can see each other but who, from isolation or some other cause cannot see anything else, we can say that they are equivalent. All the observers can determine is that the distance between them is changing at some rate. Each is just as entitled as the other to consider himself at rest and the other as moving. However, it is clear that while both may be in motion relative to the universal standard of rest, both cannot possibly be at rest with respect to it. This ultimate standard of rest is called absolute space.

Since a state of rest has duration and motion means that the moving body changes position with time, we have also to think about time. Just as with space, we experience measurable relative times. Relative times are measured by counting events which regularly recur, like the ticks of a clock, on which the hands serve only to relieve the tedium of counting by indicating the number of ticks since we last looked at it. The daily and annual rotations of the earth also serve as clocks working on longer unit intervals. In fact any regularly recurring phenomenon can be used as a clock but the recurrence must be regular, otherwise while we could tell that a certain event had happened later than another, we could not tell how much later.

The apparently simple idea of regular recurrence involves a difficulty. Suppose that we want to check that the speed of a moving body is uniform, meaning that it covers equal distances in equal intervals of time. Equal intervals of time can only be determined by something which moves or changes regularly, that is in equal intervals of time which can only be determined by . . . . . and so on. Time as we measure it will always be involved in this infinite regress or argument in a circle, depending on how you choose to regard it. Accordingly Newton had to find an ultimate standard of time to which relative time could be referred. For things to exist at all they must have duration. The universe as a whole has duration regardless of the succession of local events in any particular region and it is to this that Newton appealed as the ultimate standard of time.

To these two ultimates Newton gave the names of Absolute Space and Absolute Time. Absolue time Newton specified in these terms: — ([7])

> "Absolute, true and mathematical time, of itself and from its own nature, flows equably without relation to anything external, and by another name is called 'duration'; relative, apparent and common time is some sensible and external (whether accurate or unequable) measure of duration by means of motion, which is commonly used instead of true time, such as an hour, a day, a month, a year."

Still involved in the tautology of relative time, Newton believed that corrections could be found, as by astronomical observation, by which relative time could be so adjusted as to agree with absolute time. We must not try to imagine a cosmic clock ticking through all eternity; absolute time has no intervals and is forever inaccessible to us. Later we shall see that, except in a certain specialised sense, it is impossible to have a cosmic time for the whole universe. We no longer try to conceive of time as mere duration without a succession of events, but absolute time was thought of as flowing steadily from the infinite past to the infinite future, regardless of events, regardless even of the existence of the material universe.

For Newton absolute space was not just extension nor a purely mathematical concept like a set of co-ordinates. The heritage of the past was too strong for him to take such an abstract view, so absolute space had to be a physical reality, not the unthinkable void. As the container of matter it could be pre-existent to and independent of the material universe, but, on religious grounds, it could not stand in that relationship to God and so had to be a divine attribute. ([8]) His statement on this is to be found in the "Opticks," ([9]) but the specification in the "Principia" is severely technical.

---

7. MUNITZ, Loc. Cit. P. 202.
8. Newton's own term "Sensorium of God" was the origin of a dispute with Leibniz in which Clarke spoke for Newton. See JAMMER, M., *Concepts of Space*, Harper Torchbooks T.B. S 33 P. 112 f.
9. NEWTON, I., *Opticks*, Dover reprint S 205, P. 403.

"Absolute space in its own nature, without relation to anything external, remains always similar and immovable. Relative space is some movable dimension or measure of the absolute spaces, which our senses determine by its position to bodies and which is commonly taken for immovable space; such is the dimension of a subterraneous, an aerial or celestial space, determined by its position in respect of the earth."

Like absolute time, absolute space is impossible to be demonstrated by observation. Newton did believe that its existence could be proved by proving the reality of absolute motion necessarily involving forces, that is accelerated motion. Relative motion of a body can always occur without the application of forces within the frame of reference; it is only necessary to endow the frame with a suitable motion to achieve this. To quote Newton, in translation, on this point: —

"The causes by which true and relative motions are distinguished one from the other, are the forces impressed upon bodies to generate motion. True motion is neither generated nor altered, but by some force impressed upon the body moved; but relative motion may be generated or altered without any force impressed upon the body. For it is sufficient only to impress some force on other bodies with which the former is compared, that by their giving way, that relation may be changed, in which the relative rest or motion of this other body did consist . . . . .

"The effects which distinguish absolute from relative motion are, the forces of receding from the axis of circular motion. For there are no such forces in a circular motion purely relative, but in a true and absolute circular motion, they are greater or less according to the quantity of the motion. . . .

"It is indeed a matter of great difficulty to discover, and effectually to distinguish, the true motions of particular bodies from the apparent; because the parts of that immovable space, in which those motions are performed, do by no means come under the observation of our senses. Yet the thing is not altogether desperate; for we have some arguments to guide us, partly from the apparent motions, which are the differences of the true motions; partly from the forces, which are the causes and effects of the true motions." [11]

---

10. MUNITZ, Loc. Cit. P. 202 f.
11. From JAMMER, M., Loc. Cit. P. 103.

This idea leads to the well known experiment with the bucket of water, where the effects of "centrifugal force" are taken as evidence of absolute motion and hence of the reality of absolute space. When the bucket is made to spin there is relative motion between it and the water until the water gains a speed of rotation equal to that of the bucket, which can then be stopped and for a while the water continues to rotate with its surface strongly curved. It makes no difference to this force whether at this stage the bucket is spinning or not; hence rotation must be absolute motion.

The truth of this proposition was strongly denied by contemporary thinkers. Berkely, Leibniz and Huygens all insisted that there is only relative motion and absolute space is a fiction. Berkeley's argument, of greater cosmological appeal than the others, is that the water rotates relative to the material universe at large, represented by the system of fixed stars. [12] There can be no proof that the centrifugal forces would arise if the universe were empty but for the spinning mass of water. In fact the idea of rotation, just as of any other motion in an otherwise empty universe would be completely meaningless. Presumably just the same "centrifugal forces" would arise could the water be kept still and the rest of the universe made to revolve about it. Berkeley's argument was sound enough but the overwhelming success of Newton's mechanics nullified the effects of the criticism of Berkeley and the others as to the logical foundations of the concept of absolute space.

One may wonder why Newton did not adopt the system of fixed stars as representing the immobility of absolute space, because in his day they were really thought of as fixed in space and Digges' hypothesis that their distribution extended throughout all space was well known. However, Newton was also deeply concerned with gravitation and he conceived the part of the universe really at rest to be its centre, the point to which all bodies gravitate most and which accordingly could, least of all, have motion. Newton, on a basis of thought derived from antiquity, nominated as the centre of the universe that point from which the sun moves least, realising that the sun must have some motion about the common centre of gravity of itself and the system of planets. There is, however, no possibility of verifying this proposition by observation and the existence of absolute space remains beyond verification.

Real or not, absolute space is an important and influential concept.

---

12. BERKELEY, G., *Principles of Human Knowledge*, Everymans Library, No. 483, 1960 § *CX et seq*

Even today it retains for the average person some of the reality it had for Newton; it seems necessary as the place for things to be and to happen.

In the cosmological sense absolute space and time could provide room and duration in which the material universe could exist but some difficult problems arise from their infinite character. Men have argued about infinity for ages past and the last words about it have not been said; perhaps there are no last words to be said. Almost at the beginning of mathematics we come up against infinity in thinking about the natural integer numbers. The series of them is infinite; however large a number we may think of, it is always possible to add to it and think of a greater. There is no last number. To be more precise, the series of numbers is potentially infinite, which means that we can carry it as far as we please but we do not try to think of a last number and still less of the sum of all numbers. There is no great difficulty here, but considering fractional numbers, we have to think that between any two adjacent integers there must be an infinity of fractional numbers and between adjacent fractions still an infinity of fractional numbers. These we may wish to equate with the points of a line, of which there is an infinity in any length, to set up a one-to-one correspondence between numbers and points. Now we are faced with an actual infinity, encompassed, as it were, between finite limits. It is no longer a question of going as far as we please; once started we must go all the way. With numbers and points Georg Cantor succeeded in doing just this about a century ago, thereby starting one of the most furious of all mathematical controversies, from which rumblings can still be heard.

In cosmological time, the future may be potentially infinite, and this does not worry us much. We know no reason why the universe should cease to exist, so after any date we care to think of, we can put a later date. World without end is not a novel idea for us. But what about the past? Can we go on forever thinking of dates earlier than the one we thought of last? If we accept absolute time we can; but then if we relate time to events, to the material universe, we may have to say that the universe never did begin, for a beginning sets an ultimate date, a finite time ago. Infinite past time is actual infinity; it has been.

In terms of space the difficulty is even more severe. Newton's absolute space must be actually infinite in every direction. In Newton's day there was only Euclid's geometry which seemed self-evidently true, and space had to be Euclidean in its geometrical properties. Straight lines can be produced indefinitely but a straight line in pure geometry is not just the same as the path of an undisturbed body in a straight line. According to the first law, a moving body is able to travel along a perfectly straight line forever. To paraphrase what Lucretius had said long before, there must be place beyond place to infinity. Moreover, if Digges was right which observations with the telescope seemed to confirm, and the stars are distributed throughout infinite space, then not only is the universe infinite in extension, it must also contain an infinite quantity of matter. This must be so, however thinly the stars may be scattered and from the infinity of matter, if we can accept it all, important consequences follow.

We must now look at the cosmological consequences of the second law of motion and the law of gravitation. About these two there are several points of interest and the discussion of them begun in Newton's lifetime has not yet ended. To start with the most neutral expression of it, the second law states a quantitative relation between force, mass and acceleration. In the technical sense used here, acceleration, denoted by a means rate of change of velocity whether in magnitude or in direction [13] and rest is simply the case where the velocity is zero. The second law says that the measure of force is the product of mass and acceleration, which is expressed symbolically as $F = m\ a,$ or alternatively

$$F = \frac{d}{dt} m\ v.$$

There are several ways of looking at this law. Newton attributed a certain reality to forces, which can be illustrated by thinking of muscular action, where we can feel the application of the force. The illustration is the more apt because it seems that Newton, following his predecessors since William of Oresme and Jean Buridan, tended to think of force more in terms of impetus than of long term influences which would produce uniform accelerations indefinitely; indeed gravity is the only force of the latter kind which comes to mind in the Newtonian context. On this basis there are grounds for thinking that Newton was concerned with change of motion rather than rate of change of motion as we commonly assume today. The case for this view has been made out by Ellis on both logical and linguistic grounds. [14]

The second symbolic representation of the law, though giving the same results mathematically, has a different aspect as it overtly introduces the

13. The term deceleration for reduction of velocity is seldom used in formal scientific discourse.
14. ELLIS, B., *The Origin and Nature of Newton's Laws of Motion*, in *Beyond the Edge of Certainty*, Edited Colodny, Prentice-Hall, 1965 P. 36 f.

quantity momentum, *mv*. Force is thus measured by rate of change of momentum which is perhaps more consonant with the modern view of force, a more abstract one than prevailed in the time of Newton and his immediate successors. Momentum is a quantity obviously having a close relationship to inertia. To put a body of some mass in motion from rest, we have to overcome the inertia of its mass. In the case of a moving body which we wish to stop, we have to overcome its momentum, the product of its mass and its velocity. Coming back to the first law of motion for a moment we can see that it is a statement that momentum is conserved, remaining constant for motion in force-free space. This proposition was formulated by Huygens shortly before the first law of motion was laid down by Newton.

In the older cosmologies, the treatment of celestial motion was essentially kinematical, that is to say in terms of pure motion, happening naturally and inevitably, without forces or other influences acting to cause it. Perpetual circular motion was held to be the natural motion of the planets and needed no force to sustain it or even to cause it. When the impenetrable heavenly spheres had been discarded, the orbits proved to be elliptical and the planets taken to be of the same nature as the earth, the question arose: what keeps them in their orbits? Just as a weight can be kept in an orbit by whirling it round at the end of a cord, so the planets had to have some connection holding them to the sun otherwise they should have long since flown off into space. Among several theories proposed, one due to Descartes ([15]) involved a material connection. Like the ancients, Descartes was convinced that nature abhors a vacuum and he proposed that all space is filled with a material in which vortices form, constituting matter. On this theory the planets are entangled in the outer parts of the sun's vortex. Another theory was developed by Kepler, based on magnetism. ([16]) It was already known that the earth has magnetic properties so Kepler assumed that the sun and the other planets likewise had them. Rapid rotation of the sun would produce a magnetic vortex keeping the planets in their orbits.

Galileo studied motion under gravity ([17]) but he was more interested in the changes of motion with time rather than the forces involved and

---

15. FORBES & DYKSTERHUIS, Loc. Cit. P. 190.
16. DREYER, J.L.E., Loc. Cit. P. 394 f.
17. Galileo showed the path of projectile to be a parabola. This is true only for a trajectory short enough for the surface of the earth to be taken as flat. If the projectile be fired with enough speed to miss the earth in its subsequent fall it will go into orbit with an elliptical trajectory.

he confined most of his quantitative work to terrestrial phenomena. Whether from mere lack of interest or some stronger motive, Galileo ignored Kepler's work on planetary orbits; in consequence of these features of his approach, he was in no position to anticipate Newton's theory of universal gravitation.

Newton did not attempt a scientific theory of the nature of gravitation and was content to work on the dynamics of its operation. Newton's "Principia" was written and published in Latin, but in 1729, Motte made an English translation of the third edition of 1726, so we can read it in a version similar to the English which Newton might have used. On the nature of gravitation there is an interesting passage:— (18)

> "But hitherto I have not been able to discover the causes of these properties of gravity from the phaenomena, and I frame no hypotheses; ...... And to us it is enough that gravity really does exist, and act according to the laws which we have explained and abundantly serves to account for all the motions of the celestial bodies and of our sea."

In gravity Newton had to deal with a force capable of operating over great distances without any intervening medium and so quite unlike ordinary mechanical forces. As he says in his third letter to Richard Bentley; (19) it is absurd to suppose that gravity is an essential innate property of matter and it must be a special provision for keeping order in the universe. As a long range force gravity is unique in another way since its effect is exclusively attractive, unlike electricity and magnetism which exhibit both attraction and repulsion according to polarity. Within the limits of experimental accuracy, the attractive force of gravity is found to exert itself equally in all directions; in modern language, the gravitational field of a body is isotropic. While any two bodies attract each other according to a certain law relating the attraction to their masses and distance apart, their gravitational fields do not influence each other as do two electric or magnetic fields. These considerations will reinforce for us Newton's feeling that gravity is something very special compared with the other phenomena associated with matter.

It is known that the strength of the gravitational field around a body depends on the mass of the body and being isotropic in space of three dimensions, it decreases in proportion to the square of the distance from the body. This inverse square law for gravity has been used as an argu-

---

18. From SHAPLEY & HOWARTH, Loc. Cit. P. 77. A revised translation is in MUNITZ, Loc. Cit. P. 210 f.
19. MUNITZ, Loc. Cit. P. 217.

ment that physical space is necessarily three dimensional. [20] Newton, however, established the inverse square law from a study of the motion of the moon based on Flamsteed's observations and its orbit being elliptical according to Kepler. He was able to show finally that with gravity as the controlling factor no other orbit is possible.

The result was the law of universal gravitation saying that any two bodies in the universe are attracted to each other with a force directly proportional to the product of the masses and inversely proportional to the square of the distance between them. Symbolically, $F = m_1 m_2/d^2$. With his theory of gravitation Newton could explain the dynamics of the solar system and particularly he could show that the planets would remain in their orbits almost indefinitely. It was obvious to extrapolate the law to the universe of the fixed stars but immediately there is a difficulty. Those immobile stars had to keep their positions indefinitely even though, because of their gravity, they mutually attract one another. If the system of stars was finite, they would quite certainly collapse together around the centre of gravity of the whole. But this was an infinite system. Bentley suggested that since every star will have infinite mass around it in every direction the force of gravity which it experiences will though infinite, be balanced, resulting in stability. Newton [21] answered that it is not legitimate to equate infinities, to which the ordinary idea of equality cannot apply; as he put it — "They are neither equal nor unequal." From an earlier letter to Bentley, [22] it appears that Newton contemplated an original uniform distribution of matter through infinite space which, because of gravity, collapsed into an infinite number of large masses, the stars, separated by vast distances. How stability could be thus achieved, he does not say but perhaps his thought was carried over to the end of the paragraph where, referring to the difference between the sun and the planets, he says: " . . . . I do not think explicable by mere natural causes, but am forced to ascribe it to the counsel and contrivance of a voluntary Agent."

From observation it is clear that in our part of the universe the force of gravity is not infinitely great, in fact it is rather small, especially when compared with some of the magnitudes of other forces which we encounter. An impression of this can be gained from the low value of the

20. Whitrow, G.J., *The Structure and Evolution of the Universe*, Hutchinson 1961, P. 199 f.
21. MUNITZ, Loc. Cit. P. 215.
22. MUNITZ, Loc. Cit. P. 211.

universal constant of gravitation which is used to obtain definite values of gravitational force from Newton's third law for given masses and distance. ([23]) We now know that the only possible condition for stability in an infinite static system is the embarrassing one that the universe shall be empty. ([24]) Even the closed space finite model developed by Einstein can only have a conditional sort of stability which the slightest disturbance would upset. However, it was not realised in former times that Newton's universe could not be stable and even though there were serious doubts about it, the model was accepted because at the time there seemed to be no possible alternative. The overwhelming success of Newtonian dynamics made men shut their eyes to any argument to the contrary and there was little inducement to quibble over gravity which was so mysterious in its nature that some appeal had to be made in any case to powers beyond the grasp of science.]

In view of the importance of the infinite static universe of Newtonian cosmology with its uniform distribution of stars we should next try to gain some idea of the scale of things in the model and how far astronomers thought they had succeeded in exploring the depths of space up to the beginning of the 19th century.

23. The universal constant of gravitation, G, is 6.670×10^{-8} dynes/g/cm^2.
24. MILNE E.A., *Relativity, Gravitation and World Structure*, Oxford 1935, sec. 60.

Up from the Earth's Centre through the seventh Gate
I rose and on the Throne of Saturn sate.

CHAPTER 5

THE SCALE OF THE UNIVERSE IN EARLIER TIMES

In antiquity all that could be said about the distances of the celestial bodies was that they were very large compared with distances on the earth and thus with the size of the earth itself. Indeed the size of the earth was not accurately known until, rather late in the history of Greek astronomy, Aristotle estimated the diameter of the earth as 12,400 miles or 20,000 kilometres and this was the best figure until Eratosthenes, about 240 B.C., came very close to the modern value.

The ordinary way to measure a length or distance is to compare it with some kind of ruler, thus for distance over level ground, a stick òr chain will serve fairly well. Where the object whose distance is to be measured is inconvenient or impossible of access the method of triangulation is used. From the ends of a base line of accurately known length, the bearings of the object are measured; now, knowing one side and two angles of the triangle, it can be solved and the distance found. In astronomical practice, for finding the distance of stars, the longest possible baseline is essential and that is the diameter of the earth's orbit, some 186,000,000 miles or 300,000,000 kilometres. Even with so great a baseline the angles are extremely little different from right angles and a modification of the method must be employed. If we look at an object in the middle distance and view it with each eye alternately, it appears to change its position against the background. The true distance of the object is along the line from it to the centre of the baseline, in this case, half the distance between the eyes. By taking half the measured angle and half the baseline, we have a right angle triangle which is easily solved.

In the same way, when the angular position of a star is measured from opposite ends of the diameter of the orbit of the earth it should be found to have changed, that is, in relation to still more distant stars. Generally the word parallax is used for the whole angular shift but, because of the right angle technique, astronomers always use it for half the angle.

For this and other measurements, the radius of the earth's orbit is a very important quantity and it is known as the Astronomical Unit. To find the distance of the sun is a difficult task, just how difficult without

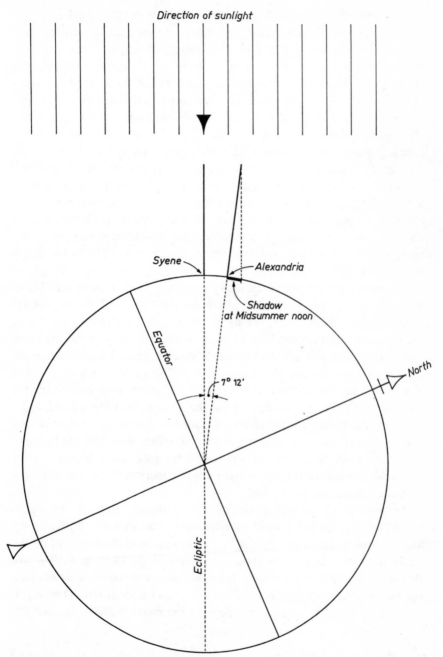

Fig. 1   Finding the diameter of the Earth.

modern instruments is hard for us to realise. Without optical aids, the blinding glare of the sun makes it impossible to measure its parallax over a terrestrial baseline with sufficient accuracy. Moreover, because of the rather rapid motion of the sun, it is important to know accurately the difference of longitude, or time difference between the ends of the baseline, or to be able to arrange in some way for exactly simultaneous observations. As Halley showed in 1716, a better approach is to take advantage of a transit of Venus across the sun. The parallax of Venus is large, almost a minute of arc, and is easy to measure when the planet is seen dark against the sun. The time of transit of Venus across the face of the sun must also be measured and then a calculation based on proportion will yield the solar parallax. Like eclipses, transits are rare events and can only be seen from a limited part of the earth on each occasion. It was the transit of Venus which Halley had predicted would be visible from Tahiti in 1769 that led to Captain Cook's expedition to the South Seas in that year.

One of the steps made in antiquity to obtain a real idea of the size of the universe was the work of Eratosthenes ([1]) who found the circumference of the earth. At Syene, in Egypt, at noon on a mid-summer day, a vertical stick casts no shadow, while at Alexandria at noon on the same day Eratosthenes was able to measure a shadow which indicated an angular distance between the two towns of one fiftieth part of a circle, $7°12'$. Fortunately for him the terrain along the Nile is fairly level and Eratosthenes was able to get a good measure of the distance over the ground, some 5,000 stadia, giving a value of 250,000 for the circumference. Converting this to modern units, he was only about 240 miles, some 380 km in error. Whether or not Eratosthenes was aided by good luck, it was a very remarkable achievement considering the comparatively primitive means of measurement then available.

Another step was the determination of the distance of the moon according to a method devised by Hipparchus and Ptolemy. ([2]) Referring to Fig. 2, we have an observer at 0, on the surface of the earth, while the moon is moving in a circle about the centre of the earth at C. The arc through which the moon travels between rising and setting is determined by the observer's horizon and is clearly less than a semicircle. How much less can be found by comparing the time the moon is above the horizon

1. DREYER, Loc. Cit. P. 174 ff.
2. For another method used by these and later astronomers, see DREYER, Loc. Cit. P. 183.

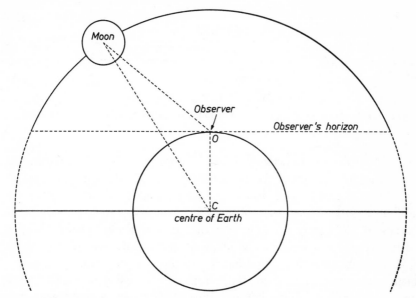

Fig. 2   Distance of the moon from the earth (Hipparchus and Ptolemy).

with the time it is below. Knowing the radius of the earth from Eratosthenes, it is possible, either by geometrical construction or by calculation, to find the distance of the moon. With modern equipment this can be determined as a little over 60 earth radii. The value obtained by Hipparchus was between 67 and 78 earth radii. but it was not until the 16th century that a better figure was given by Tycho Brahe who worked on the lunar parallax and made the distance 233,000 miles or 375,000 km, only a little less than the present value.

In terms of the distance of the moon, the distance of the sun can be found by a method due to Aristarchus of Samos, [3] indicated in Fig. 3. At the moment when the moon is exactly half sunlit and half dark as seen by an observer on the earth, the line from the centre of the sun to the centre of the moon must be at right angles to the line from the centre of the moon to the observer. If we now measure the angle between the centres of the sun and the moon at the observer's position, we can find the ratio of the distances of the sun and the moon from the observer. Aristarchus obtained for this angle a value of 87° from which he con-

3. DREYER, Loc. Cit. P. 136.

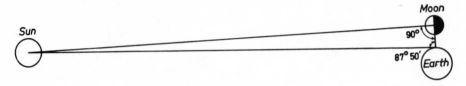

Fig. 3   Distance of the sun in term of distance of the moon (Aristarchus of Samos).

cluded that the sun is between 18 and 19 times as far off as the moon. The value for the actual distance of the sun is 87°50′, corresponding to something like 400 moon distances. Although his result was numerically very much in error, the attempt made by Aristarchus was very important because it was a truly scientific approach on a principle which could give correct results and lacked only adequate instruments.

It was not until the 19th century that precise telescopic observation made it possible to measure the parallax of a nearby star and in the meantime attempts were made by other methods which could give only approximate results compared with parallax. In earlier times it was not possible to detect any stellar parallax at all but this made it clear that the stars were much more distant than the planets and that the whole system was so vast in extent that the size of the earth was insignificant. As Copernicus put it: —

> "For what was proved was only the vast size of the heavens
> compared with the earth, but how far this immensity extends is
> quite unknown."

In the 17th century we find Huygens (4) deploring as unfounded speculation the idea of Kepler that as the orbit of Saturn is 2,000 times the diameter of the sun, the sphere of stars ought to be 2,000 times greater in diameter than the orbit of Saturn. Tycho Brahe stated as one of his objections to the Copernican theory that the stars would have to be impossibly large on that model. If we make naked eye measurements of the apparent size of a star, optical effects lead to measurements around 3′ of arc which means an enormous physical size at a great distance. However Huygens had made a major contribution to the design of telescopes and was an expert in their use, so he knew that in a telescope stars appear as points of light having no appreciable angular size. Christiaan Huygens, who was probably better known to his contempo-

---

4. MUNITZ, Loc. Cit. P. 221.

raries as Mynheer van Zulichem, a title deriving from his family estates in the Netherlands, was one of the leading scientific figures of the 17th century and performed as brilliantly in other fields as in optics.

In an attempt to make an indirect measurement, Huygens fitted to a telescope an opaque mask in which he could make very small holes so that he could see a tiny portion of the sun's disc. Trying progressively smaller pinholes, he finally achieved one which appeared to give an image with the same brightness as Sirius. According to his calculations, the image was that of one 27664th part of the surface of the sun and so the star would be at 27664 times the distance of the sun. By the same method Huygens could calculate the angular size of the star which came out at 4‴, showing that any attempt to measure the angular size would be quite beyond contemporary technical resources. (⁵) Obviously the method which Huygens used to find the distances of stars involved some very serious subjective difficulties since the observations of the sun and Sirius had to be made some hours apart, which meant that he had to estimate the brightness of one in terms of his memory of the other. Only photography or some other way of providing an objective carry-over of the brightness information from one observation to the next would offer any reasonable chance of success. It is not surprising, then, that Huygens' measure for the distance of Sirius was a long way out, the true measure being over half a million times the distance of the sun.

At the very beginning of the 19th century we find H. W. M. Olbers, who will figure later in our studies in another connection, trying Huygens' method with the then recently developed technique of photometry, which does give a fairly objective transfer of brightness. With the aid of suitable optical devices it was now possible to observe simultaneously the images of the star and of a local light source which could be adjusted in brightness to coincide with the star image and by means of a calibrated scale the same brightness could be found again for the next observation. Olbers worked in two steps, using a planet as the intermediate between the sun and the star. In this way he found that sunlight is about 120,000 million times as bright as a first magnitude star which would have an angular diameter about 0″.003; (in the same notation Huygens' figure is 0″.06). The resulting distance is about 312,000 A.U. About the same time Sir William Herschel approached the problem in a similar way obtaining a result of 344,000 A.U.

---

5. The notation 4‴ is an old form in which the ‴ means the 60th part of a second and was called a Third.

The method of brightness comparison, as employed by these astronomers, was based on an assumption, generally adopted in their times, that all stars are of much the same intrinsic brightness and that the differences in apparent brightness are the result of different distances in accordance with the inverse square law of the propagation of light. Accordingly, once it was felt that the distance of nearby stars had been established the distances of the fainter stars could readily be found. When the precision of instruments became great enough for the method of parallax to give useful results it was natural to select the brightest stars as necessarily the nearest but this did not work out as expected. As we now know, there are enormous differences in intrinsic brightness between the stars.

When Halley, a great astronomer and younger friend of Newton, made a new catalogue of the positions of a number of stars he found that some of these differed significantly from the positions given by Hipparchus and Ptolemy nearly two thousand years before. [6] As some of the discrepancies were well beyond the limits of accuracy to which Hipparchus had worked the only conclusion to be reached was that the stars are not absolutely fixed but have indeed proper motions of their own, these however, being insignificant by comparison with the vast distances between the stars. Further and more precise observations showed that some stars have proper motions large enough for the change in position to be detectable in a period of a few years. Regardless of the apparent brightness, it now seemed eminently reasonable to assume that these stars with the greatest proper motion would be the nearest, an assumption which has proved correct.

In actual application the method of parallax involves dealing with several factors apart from any change of position caused by proper motion. One, the precession of the equinoxes, had long been known as the reason for a shift of the constellations along the ecliptic of rather more than half a minute of arc a year and a complete circuit in 25,000 years. In the 18th century, Bradley provided the measures of two other effects, of which the first was the aberration of light, [7] a consequence of its finite velocity. At opposite points of the earth's orbit its motion is in opposite direction and its speed, thus relatively doubled, makes the light of the stars appear to come from a slightly different direction, just as the direction in which rain falls seems to change according to one's

6. HALLEY, E., *The Detection of Proper Motions,* Phil. Trans. 1717, reprinted in SHAPLEY & HOWARTH, Loc. Cit., P. 101 ff.
7. SHAPLEY & HOWARTH, Loc. Cit., P. 103.

speed and direction of walking. The astronomical result is a very small but important annual change in the observed position of the stars. Secondly, the axis of the earth has a slight wobble, technically called nutation, causing another positional change with a period of about 18 years.

When making parallax observations astronomers make their measures of position of the selected star with reference to stars which have small angular separations from it and the smallest proper motions. In the course of determining the positions of such reference stars Herschel and Struve found large numbers of close pairs of stars, many of which turned out not to be merely close in respect of angular position but to be genuine pairs gravitationally bound and moving in orbit around each other. Before any real attempt could be made to measure parallaxes many years of these preliminary observations were required and, to ensure reliability, the final task itself took several years to accomplish.

Finally in a letter to Sir John Herschel in 1838, F. W. Bessell [8] was able to announce that as the result of observation made since 1834, he had found the parallax of the star 61 Cygni for which he gave the figure 0".316. This makes the distance of 61 Cygni no less than 657,700 A.U. a distance which it would take a ray of light 10.3 years to traverse. Very shortly after this the parallaxes of two more nearby stars were published, that of α Lyrae by Struve and that of α Centauri by Henderson. At a distance of 4.3 Light-years α Centauri is the second nearest star to the sun; only the faint star Proxima Centauri is a little nearer.

In the preceding paragraphs we have used several astronomical terms which require some explanation as they may be used later on in contexts where their meaning is important. The term light-year came into use once the measurements of star distances took us far outside the solar system and distances became so great that when expressed in miles or even in A.U. the numbers are too big to be handled conveniently. At 186,000 miles or 300,000 km per second, the distance travelled by light in a year was found to be a useful unit for stellar distances. Approximately it is 66,000 A.U. As more powerful telescopes came into use the scale of distance increased enormously and we now know of objects whose distances have to be stated in thousands of millions of light-years. An alternative measure of distance is the parsec, the distance at which a star would have a parallax of one second of arc, that is about 3.26 light-years.

---

8. Ibid, P. 216.

For objects very far off there are two multiples in common use, the kiloparsec and the megaparsec.

The word magnitude also has a definite technical significance. Apparent magnitude is the measure of brightness of a distant body as seen from here. Magnitudes are reckoned on a logarithmic scale, the difference in brightness for one magnitude being 2.512 times. Thus a star of the first magnitude is 100 times as bright as one of the 6th magnitude. The starting point of the scale was settled long ago as the brightness of the very brightest stars we can see, the first magnitude. Obviously apparent magnitude can give no indication of the intrinsic brightness of a star because of the differences in both brightness and distance. Once the distances of stars could be established it was possible to adopt a standard called absolute magnitude which is the brightness which the stars would have if seen from the same distance which has been set at 10 parsecs. The sun has an apparent magnitude of —27 because of its nearness to us but its absolute magnitude is only 4.9, not a very bright star at all. The brightest known star is S Doradus in the larger Magellanic Cloud at —8.9 mag. absolute.

Towards the end of the 18th century the elder Herschel had surveyed the northern sky and catalogued several millions of stars and had plumbed the depths of space to about 500 times the distance of Sirius on his own reckoning, thus approximately 5,000 light-years. For a long time Herschel accepted the Newtonian scheme of stars uniformly distributed to infinity but in the end his own observations led him to change his mind. Indeed F.G.W. Struve was the only great astronomer who held to the theory of uniformity until the middle of the 19th Century. By 1785 Herschel was seeing the Milky Way as a rather local system of stars and working on the theory that the nebulae are similar star systems scattered throughout infinite space. This idea can fairly be called uniformity on the larger scale.

Supported by the great authority of Herschel and the speculations of Wright and Kant, the new theory had won some acceptance by the early 1800's. Now unlike the stars, planets and comets the nebulae were gaining cosmic significance. But it was too early yet for a model with an endless distribution of galaxies to be viable. Whether it was taken to be composed of galaxies or of stars in uniform distribution, the infinite static universe involved an insuperable difficulty. Once we see what that difficulty was, it becomes so obvious that we wonder why it took so long to be appreciated and our next task is to look into the circumstances surrounding it.

CHAPTER 6

THE PARADOX

The main features of the Newtonian model of the universe or as it is also called, the classical model, have now been established and may be conveniently summarised as follows: —

(a)   The stars are all similar to the sun in size and intrinsic brightness.

(b)   The differing apparent magnitudes of the stars are simply the consequence of their differing distances from us in accordance with the inverse square law for the propagation of light.

(c)   The universe is homogeneous and isotropic.

(d)   On the large scale the universe is static.

(e)   The universe is infinite in extension.

Although it is not relevant to our present purpose, it was always assumed that space is Euclidean, indeed this was taken to be a self-evident truth.

The first four of these statements were supported by observation, at least until the results of Herschel's great surveys became known. The fifth statement appeared consistent with visual observation and because of its intimate connection with the very foundations of Newtonian mechanics, had obviously to be adopted.

As to the third statement, we should note that homogeneity and isotropy are not just two names for the same idea though they are closely related. As an indication of the relations between them we can say that in most cases homogeneity depends upon the properties of matter while isotropy depends on its spatial distribution and motion. Taking the example of a beach, we may say that the sand is homogeneous if a handful taken from one part of the beach is very like a handful taken from any other part. But a beach is not isotropic because its characteristics are not the same in all directions; thus it will be wetter in one direction and dryer in another. The Newtonian universe is homogeneous because the number of similar stars per unit volume of space is pretty much the same everywhere, provided of course, that we choose a large enough unit of volume. Isotropy follows from there being no difference on the large scale in the properties of the universe in any direction. If we were to find that the number of stars per unit volume decreased in the same way with distance from us in every direction the universe would still appear isotropic but certainly not homogeneous in the ordinary sense of the word. The word static is used here in a technical sense, meaning that there are no large scale systematic motions in the universe. The scale is such that we are not interested in the orbital motions of the planets around the sun; we can also allow the sun and other stars each to have a small random proper motion of its own. But these motions must not, even over long periods, of time, upset the uniform distribution of stars in space. A systematic motion is one common to all the stars and, whether a motion of translation or of rotation, would be definable in Newtonian terms as a general motion relative to absolute space. Presumably Bishop Berkeley would have denied the meaningfulness of such a definition.

It is important to recognise that the five statements about the universe set out above are all assumptions or hypotheses. Not one of them is a statement of fact guaranteed in any way. They are hypotheses derived either immediately from theory, as in the case of (e), or they are based on observations which in turn have some theoretical background. It really does seem extremely doubtful that anyone ever makes a pure observation; there is always something more reported than the bare facts. When someone looks at the sky and reports that he can see a star there is a substantial body of theory involved in that apparently simple statement. Very often we are not consciously aware of all the assumptions used in normal scientific discourse and if we were to try and bring them all out it would effectively stifle our attempts to say anything. Sometimes hidden assumptions are both important and very difficult to detect but in both science and pure mathematics the effort has, on occasion, proved highly rewarding.

In the case of the set of assumptions with which we are now concerned, taken one by one, they all seemed to be quite acceptable in terms of contemporary theory and observation and it appeared reasonable to suppose that they could be put together to make a workable theory of the universe. Newton and his contemporaries did not realise that they had created a problem and it was not proved until afterwards that something was radically wrong with the model. To misquote an old adage, this was a case where the whole was worse than the sum of the parts.

It is now well known that, in 1826, H. W. M. Olbers, [1] then directing

---

1. OLBERS, H.W.M., *Ueber die Durchsichtigkeit des Weltraums*. Bodes Jahrbuch, 1826, P. 110 ff. Reprinted as Appendix I of this book.

The 210 foot radio telescope at Parkes, New South Wales with which many important discoveries have been made. Photograph by courtesy of the Radiophysics Division, C.S.I.R.O. Sydney. Weighing 450 tons this telescope can be steered with an accuracy as fine as 1 minute of arc.

A classical telescope. Originally built in 1868 as a 48 inch reflector for the Melbourne Observatory this instrument has been re-erected at Mount Stromlo Observatory, fitted with a 50 inch mirror and provided with modern electronic control. The original speculum metal mirror can be seen mounted on the wall.

the observatory at Bremen, published a paper presenting a simple observational proof that the Newtonian model was wrong. More recently it has
become known that the Swiss astronomer, J. P. L. de Cheseaux ([2]) said
the same thing in almost the same terms in 1744. ([3]) According to Olbers,
Halley had also been doubtful of the model for the same basic reason
but through one of those simple mathematical errors which now and then
trap great scientists, he thought his doubts were mistaken ([4]), ([5]). As a
point of historical interest, Cheseaux published his treatment of the
problem as an appendix to a book on another topic and it did not attract
much notice. Olbers actually had a copy of the book in his library but
we have the evidence of F. G. W. Struve ([6]) that he had never read it and
so must be credited with the independent discovery of the proposition
that has become known as Olbers' Paradox. As the texts of these papers
are not easily accessible to students both are reprinted as appendices in
this book.

The essence of the paradox is an observation that anyone can make
and probably has been made by almost everyone without realising it:
the sky is dark at night.

There are several lines of thought along which we may work to find
out how this astonishing result follows from the apparently simple and
admissible set of assumptions on which the model was based. The sun,
which we see as having an angular diameter of half a degree, is the only
star that occupies a sizeable area of our sky but, however small the areas
occupied by the other stars may be, they are still of some finite size. It

2. CHESEAUX, J.P.L. DE, *Traité de la Comète*, Lausanne 1744, P. 223 ff.
   Reprinted as Appendix II of this book.
3. The work of Cheseaux first came to my notice in 1960, when, after hearing
   Prof. H. Bondi talking about Olbers' Paradox, I wished to find out what his
   contemporaries and successors had thought about it. A clue to Cheseaux was
   found in P. G. Tait's article on light in the Encyclopaedia Britannica, 9th
   Ed., and I procured a photocopy of the relevant pages of Cheseaux's book
   through the good offices of Prof. H.B.G. Casimir. In 1961 my correspondence
   with the late Otto Struve, following his review of Bondi's "Cosmology" in
   Physics Today, led to his sending me a copy of F.G.W. Struve's "Etudes
   d'Astronomie Stellaire", containing comments on Cheseaux, in exchange for a
   reproduction of my copy of Cheseaux, which he had not previously seen.
   Shortly afterwards I was also able to hand Prof. Bondi another copy of
   Cheseaux, of whom he had not then known. For biographical details and other
   information about Cheseaux I am gratefull for references subsequent uncovered
   by Prof. B. J. Bok, then Director of Mt. Stromlo Observatory, Prof. W.
   Becker, Director of the Observatory at Basel, Switzerland, and Dr. G. A.
   Tammann of Mt. Wilson.
4. OLBERS, H. W. M., Loc. Cit. P. 112.
5. NORTH, J. D., Loc. Cit. footnote to P. 18.
6. STRUVE, F. G. W., *Etudes d'Astronomie Stellaire*, 1847 P. 84.

follows that if there is an infinite number of stars evenly distributed in
space, the whole finite area of the sky must be covered by stars, for which
a finite number will suffice. The sky ought to be as bright as the sun
over its whole extent at all times. (7)

   Both Cheseaux and Olbers based their approaches on the consideration

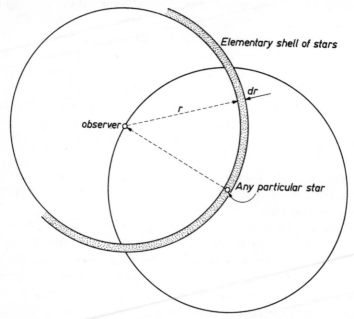

Fig. 4. The Paradox: the number of stars increases as r^2 but the light received from
each star decreases as r^2 so that the contribution of light from a shell is independent
of r.

of spherical shells of space containing stars, (8) the thickness of the shells
being very small in comparison to the radius, like gigantic onion skins.
Such a shell is a region of space bounded by inner and outer spherical
surfaces. When the thickness is very small compared with the radius of
either surface, it is quite accurate enough to take its volume as the product
of its thickness and the area of either surface: the volume of the shell is
then seen to be proportional to the square of the radius. As the number
of stars per unit volume is constant, the number of stars in a shell is also
proportional to the square of the radius. In consequence of the law of the

7. SCIAMA, D. W., *The Unity of the Universe*, Faber, 1959, Chapter 6.
8. BONDI, H., Loc. Cit. Chapter 3 follows this approach.

propagation of light, the quantity of light received from each star in a shell is inversely proportional to the square of the radius. These dependences on the square of the radius balance each other, with the result that the total quantity of light received from shells of different radii is the same.

Now in an infinite universe, we can take as many shells of increasing radius as we please, each contributing the same amount of light and it is obvious that only a finite number of such shells will provide complete and uniform illumination of the sky. We need go no further out with the shells than the distance at which the light from still more distant stars is blocked by the stars in nearer shells and the paradox will arise. It follows that for a uniform static model with the stars all similar the universe is not only not infinite, it must be no larger than certain finite calculable size.

Neither of our authors was willing to abandon any of the bases for the model; Olbers, indeed, went beyond purely scientific considerations and appealed to theology. The following is a rough translation of what he wrote: —

"But has the penetrating sight of the immortal Herschel reached near, or only somewhat nearer to the boundaries of the universe? Who can believe that? Is not space infinite? Can limits to it be thought of? And is it conceivable that the almighty Creator would have left infinite space empty?"

For both our authors the day was long since gone when one could think about finite space: the uncompromising success of Newtonian dynamics had made a physical reality of Euclid's geometry. When the infinity of space cannot be questioned it becomes very difficult for believers in the uniform distribution of matter to turn to concepts involving a finite material system. There is always in the background the problem of a centre of gravity in the system. Though one may appeal to projectile forces, as centrifugal forces were then called, it is not a simple affair to set the material universe rotating: there ought to be observational evidence of such a condition, but there clearly was none. To both Cheseaux and Olbers the idea of a rather small material system floating in infinite space was unacceptable. One can realise the repugnance which Cheseaux felt for the conclusion to which he was being driven. "Not only must the universe be not infinite, but incomparably smaller than the first shell which I have supposed."

In a situation where the set of assumptions leads inevitably to a contradiction with an observed fact, one or more of the assumptions must be sacrificed or some additional hypothesis has to be brought in *ad hoc* to save the situation.

In this particular case Cheseaux and Olbers, quite independently, elected to modify the law of propagation of light, which obviously leaves the fundamentally appealing features of the model untouched. Further it was possible to make the desired modification without upsetting the important and highly verifiable feature of the inverse square law, that light travels in straight lines. All that seemed necessary was to allow the apparent magnitudes of stars to change, not precisely according to the inverse square of distance, but at a very slightly faster rate. Such a small deviation from the simple law would be quite negligible on the small scale but would have important consequences for large distances: those faint stars are not as far off as you think.

An apparently simple way to bring about such a modification of the transmission of light without interfering with its geometrical aspect was to suppose that space is not perfectly transparent. Of course Newton's absolute space could not have such a defect, but there were enough assorted ether theories known that it was rather obvious to suggest the hypothesis that space is occupied by some medium too rare to affect the motions of bodies and which, not being perfectly transparent, absorbs some small proportion of the light which passes through it. We should not take if for granted that this medium is the luminiferous ether of classical physics. In the days of Cheseaux and even of Olbers, the corpuscular theory of light prevailed and the ether wave theory was yet to come. F. G. W. Struve, writing in the scholarly Latin still often used in the early 19th century, referred to this hypothetical medium in the words: —

> "*Si vero spatium inter stellas fixas interjacens materia aliqua non absolute pellucida, quamvis longe subtilissima, est implimentum, lumen in transitu ab altera stella ad alteram absorbatur necesse est.*"

Note the words *materia aliqua,* meaning some material, but nothing as specific as luminiferous ether.

According to Cheseaux, it would provide sufficient absorption if the medium were 33×10^{16} times as transparent as pure water. Olbers thought he needed only enough absorption to diminish the brightness of first magnitude stars by one part in eight hundred. On this basis Olbers considered that the light from stars more than 30,000 times as distant as Sirius would be completely absorbed. This degree of absorption would leave the universe infinite but limit the observable portion to a sphere about 300,000 light-years in radius. In passing we may note that this does not involve any inconsistency with the cosmological principle.

In our times the outlook on ways of escaping the paradox is quite different: we are strongly opposed to the introduction of an *ad hoc* hypothesis and we have learned that most of the hypotheses underlying the Newtonian model were wrong. The only one still generally accepted is that the universe is homogeneous and isotropic.

It is now well established that the stars are by no means all similar in size and intrinsic brightness. We have gone much further in our ideas of what are the principal structural units of the universe; the place of the stars has been taken by the galaxies. Our Milky Way is one of the galaxies and, though it is typical of its class, the galaxies are known to differ greatly in size and brightness. Also like stars they have a tendency to form clusters and it may be that these clusters are important structural units. There is a curious geometrical possibility of avoiding the paradox; the idea seems to have begun with Lambert in 1761 and was developed by Charlier, (9) early in this century. It is the hierarchical model in which the galaxies form clusters and these in turn form clusters of clusters and so on. Considered on a sufficiently large scale, the volume of space occupied by the higher orders of clusters of clusters grows more rapidly than their content of matter, so that for extremely large regions of space, the average density of matter, the amount of it per unit volume, becomes steadily less with increasing volume and in the limit becomes vanishingly small. Although on the largest scale this model is sufficiently empty to avoid the paradox, the density of matter can, on the scale of the observable universe, be in agreement with what we do observe. There is some evidence now available for clusters of clusters of galaxies but there does not seem to be any for higher orders and the model is not now of much interest.

Although it is nowhere explicitly stated by Newtonian cosmologists, there is a rather timeless character in the model: no dynamical reason appears why the universe should not be very old. Newton himself had pointed out that since all motion ultimately dies away the universe as we know it must have a finite past and future existence but the time scale for this can be very long indeed compared with the time scales of history. Had astronomers been interested in the chronology of Archbishop Ussher who, from historical and biblical evidence, dated the creation of the universe in the year 4004 B.C., a ready-made escape would have been available. The sky should be dark at night because the light from stars 6000 or more light-years distant had not had time to reach the earth.

It is now generally agreed that the universe is not static. The dis-

---

9. NORTH, J. D., Loc. Cit. P. 20 f.

covery made some 50 years ago that the light from the most distant bodies is shifted towards the red end of the spectrum, finds its most satisfactory interpretation in the Doppler effect which implies that these bodies are all moving away from us at high speeds. The speed appears to be proportional to the distance and it seems that the whole universe is expanding. Here we have an example of a universal systematic motion in which Berkeley would have been interested. In a system which the observer sees as spherically symmetrical about him an expansion which involves only radial motion of recession on the part of all distant bodies has a clear meaning without any reference to absolute space. As might be expected in the face of such an astonishing notion alternative interpretations of the red shift have been sought but none has proved widely acceptable. There are other reasons, as mentioned in the previous chapter, for rejecting static models of the universe.

We have seen that the paradox arises not only for an infinite Newtonian universe but even for a finite one of rather modest dimensions. On the basis of Sciama's calculations ([10]) that most of the light causing the paradox would have come from regions in the order of 10^{17} light-years distant, the Newtonian universe would not have to be more than a hundred or so times larger in linear dimensions than the portion of the universe which we can now observe. Though the majority of modern cosmologists confine themselves to finite models, there is an important minority favouring infinite universes. Both schools of thought appear to take the same way out of the paradox which in principle will arise as well for galaxies as for stars. We take the expansion of the universe to be real and one of the consequences is that the Doppler effect so reduces the energy of the radiation from the most distant bodies that the paradox cannot arise. This proposition is one to be taken on faith at this stage but in later chapters the evidence for it will be presented and other consequences explored.

In his treatment of the paradox Bondi ([11]) has pointed out that there is an insuperable difficulty in the hypothesis introduced by Cheseaux and Olbers. It follows from thermodynamic considerations that any medium absorbing radiation in the way proposed would in turn be heated by the accumulated energy and would have to shine with a brightness like that of the stars and so restore the paradox. In the times when the absorption hypothesis was formed the study of thermodynamics had not yet taken

---

10. SCIAMA, D. W., Loc. Cit. P. 79.
11. BONDI ibid.

shape and so the answer, now rather obvious, to the question about what happens to a medium which continually absorbs energy, could not be obtained. It is exceedingly difficult to get useful answers when you do not know what are the right questions to ask.

There is another difficulty about absorption by a translucent medium, which might have occurred to scientists in 1826 and 1837 (Struve) if not to Cheseaux in 1744. It is simple enough to consider total absorption by an opaque medium, which might for our purposes consist of scattered particles. Given that each particle is opaque, a sufficiently thick, but finite cloud of them could bring about total absorption. This is precisely the reciprocal problem to that considered by Cheseaux and Olbers. But if we postulate some continuous medium which is translucent and at no point totally opaque, it cannot produce complete absorption of the light of stars at finite distances.

Let us put on for a moment the mantle of Zeno. Suppose a star at such a distance that half its light is absorbed: now take it to twice that distance, to twice that distance again and so on; some light will always get through. We can express this symbolically putting d for distance through the absorbing medium, a for the absorption coefficient and M_0 for the intrinsic brightness of the star. Then M, the apparent brightness at some distance is given by the equation

$$M = M_0 \ e^{-ad},$$

where e is the base of the Neperian logarithms. This formula is one of the known exponential kind in which M will, with indefinitely increasing d, approach asymptotically to zero but will not actually become zero for any finite value of d. In other words, because the number of stars in the model is infinite and the absorption can never reduce the light of any star to zero unless it be at infinite distance, the absorption hypothesis cannot save the model.

Virtually alone among the great astronomers of his day Struve took up the absorption hypothesis. He was in fact one of the last strong adherents of uniform distribution and from this point of view absorption was highly desirable. In his statistical analysis of the numbers of stars at distances corresponding to their apparent magnitudes, Struve found results which seemed to give strong support for the absorption hypothesis and he concluded that for stars of the 5th magnitude as observed, rather more than half the light had been absorbed and two thirds for those of the

---

12. STRUVE, F. G. W., Loc. Cit.

6th. ([13]) However the overwhelming observations of Sir William Herschel and the precise determination of parallax by Bessel in 1838 made the uniformitarian position untenable so that the paradox was no longer a serious problem and the *ad hoc* absorption hypothesis no longer interesting.

Meanwhile the paradox and its resolution remain outstanding examples both of scientific thought and of the dangers of exploring the universe in terms of assumptions that are attractive but too tenaciously held in the face of observation. Still, out of it all grew a new model of the universe. In the process of its birth much beauty and grandeur of concept were lost; indeed we shall see how brilliant parents came to inflict upon cosmology a dull and dwarfish child.

13. STRUVE, F. G. W., *Stellarum Dupl et Multi Mensurae,* 1837 P. XCII ff. The copy of this book which I have used is in the library at Mt. Stromlo Observatory and it is from P. XCII that the Latin quotation in this chapter is taken.

With them the Seed of Wisdom did I sow
And with my own Hand laboured it to grow.

CHAPTER 7

THE ISLAND UNIVERSE

Although the 18th century astronomers were convinced of the necessary truth of the Newtonian theory including the uniform distribution of stars, three thinkers who worked on more speculative lines had already arrived at radically different concepts of the structure of the universe. Even to the naked eye the Milky Way was an apparent departure from uniformity but telescopes did not show any part of the sky as empty and many more instances were seen of the tendency for stars to appear in clusters. These appearances were taken to be essentially optical and it did not seem likely that stars of differing brightness could be physically associated in groups. However it was accepted that stars of similar brightness could be gravitationally associated in clusters. Still more, the power of Newtonian theory over the minds of astronomers was so great that these phenomena could be dismissed as mere appearances, not derogating from large scale uniformity. By this time quite a few nebulae had been found and some of these were taken to be localised aggregations of stars but having no particular cosmic significance. The Andromeda nebula which was the only one visible to the naked eye, seems to have been known since the 10th century. Six were listed by Halley in 1714, and, when Lacaille made observations in the southern hemisphere he discovered more than 40 new nebulae. In 1781, Messier the comet seeker, listed 103 nebulae similar in appearance to comets, but distinguishable from them by lack of motion across the sky. These nebulae were usually considered to be large clusters of stars at such enormous distances that the individual stars could not be resolved.

In 1750, Thomas Wright published a new theory of the structure of the Milky Way. [1]) He held that it was a genuine association of stars in the form of an enormous disc, hence it has become known as the "Grindstone" theory. According to Wright the galaxy extends to infinity in the plane of the Milky Way but in the direction at right angles to that

---

1. WRIGHT, T., *An Original Theory of the Universe,* 1750, Reprinted in Munitz. Loc. Cit. P. 225 ff.

plane it has a finite thickness roughly equal to twice the distance to which contemporary observations could reach. That the galaxy does not appear symmetrically placed in the sky is explained quite simply by supposing the sun to be at a large distance from the central plane of the system. Wright also considered that the stars could form real clusters. In this way he could account for the nebulae and other close clusters, for he refers to:

> " . . . . the diversity of motion that may naturally be conceived amongst the stars themselves, which may here and there, in different parts of the Heavens, occasion a cloudy knot of stars"

Immanuel Kant was impressed by Wright's model of the galaxy and extended the idea into an elaborate model of the universe infinite in all directions. (2) This great philosopher was firmly convinced of the truth of Newton's laws which he believed to reveal not just opinions, but real knowledge of Nature; (3) the human intellect could at last grasp some portion, however small, of ultimate reality. But infallibly true as he believed the laws must be, he was by no means satisfied with the accepted model of the universe based upon them. It seemed that a more or less random static scattering of stars in space was a poor tribute to the Creator and that some much grander scheme of things must prevail.

Starting from the planets revolving around the sun and keeping for vast periods of time in stable orbits by virtue of the balance between gravity and rotation, Kant felt he saw how to solve the old problem of stability in the universe at large. The galaxy must be a finite structure like a great disc, rotating about its centre. But it would not rotate as a solid body; the stars move in orbits like the planets, according to Kepler's law that the period of rotation is proportional to the square root of the cube of the distance from the centre. That no such systematic motion had been observed was readily explained since the period, even for relatively near stars, would be in the order of millions of years and the annual motion could not be detected without major improvements in technique. Kant followed Wright in taking the sun to be rather far removed from the centre of the system. Obviously its velocity would differ little from those of adjacent stars and the apparent motions of distant ones, though differing largely in linear velocity, would only show as minute annual changes of position.

The next step was to identify nebulae as structures similar to the galaxy. Though they all appeared as hazy patches of light some were definitely

2. KANT, I., *Universal Natural History and Theory of the Heavens*, 1755, Reprinted in Munitz. Loc. Cit. P. 231 ff.
3. POPPER, K.R., Loc. Cit. p. 184 ff.

circular in shape and others elliptical which could well be discs inclined at an angle to the line of sight. Some of them showed a spiral structure in their discs, good evidence of rotation. These nebulae suggested to Kant a system of the world in which they took the place of stars as the bodies which are distributed throughout space. Just as in any one galaxy stability could come about through the balance of gravity and centrifugal force due to rotation, so the system of galaxies might remain stable if it was in rotation around some centre. If this idea could be made workable the old problem which had beset Newton would be solved. At first sight an infinite universe which Kant's model necessarily had to represent, ought not to have a centre about which to rotate. Before we see how Kant could justify the assignment of a centre, we must look at his idea of how the universe comes into existence.

Along with the other Newtonian concepts Kant accepted absolute space. But in featureless, infinite absolute space there can be no centre. The universe according to Kant was not created out of absolutely nothing but out of Chaos, a formless, indeterminate, primaeval substance entirely occupying the whole of absolute space; almost chaos could have *been* that space and like it cannot have a centre. An infinite universe can only have a centre in a very special set of circumstances: it must have a material content with certain characteristics. Some characteristic of the material must change according to a definite law, progressively and in a spherically symmetrical way from some region outwards to infinity. Thus the required circumstances arise if the density of matter, having some finite value at some place, falls off asymptotically to zero in every direction according to an appropriate law. Then at any finite distance however great, the density will always have some finite value and will only become zero at infinity. In principle it would be possible, by exploring such a universe and observing the changes of density, to arrive at the region of greatest density and identify it thereby as the centre.

It was in just this way that Kant could have a centre for his infinite universe. The process of creation began at some place in the chaos, the formless undifferentiated substance there becoming changed into matter with an organised structure. The creation which Kant envisaged must not be thought of as something fully accomplished; it is a continuing process, most complete at the centre of the universe, less and less complete as the distance from the centre increases but always going on. In the region of the centre, where the greatest proportion of the formless chaos has been converted into matter and has collected into masses making stars and galaxies, the density of matter in space is higher than anywhere else.

Very far out from the centre there is much less matter so far formed and consequently fewer galaxies per unit volume of space and the density of matter falls off continuously to zero at infinity.

Far different as it is from the modern theories involving continual creation, the process described by Kant can fairly be described as continual creation. The essential difference is that in the modern theories creation is happening everywhere at the same rate and the new matter may come absolutely out of nothing; in Kant's model the rate of creation changes exponentially with distance from the centre as new matter comes into existence out of almost but not quite nothing.

As we have seen, the price of stability was systematic motion on the large scale but we ought to distinguish between this kind of non-static universe and those we contemplate today. In the Kantian model the motion is rotatory but in present day models the systematic motions are those of expansion, thus in the line of sight, not across it. Contrary to the ordinary usage of the word, expansion in cosmology does not imply occupying more of space. Since the theory of absolute space has been abandoned, space and the material universe coincide: space is where matter is. By expansion of the universe we are to understand that the distances between the galaxies, or clusters of them, are increasing with the passage of time. There is a temptation to think of the Kantian universe as expanding but this is true only in a rather special sense: what gets larger is the distance from the centre at which the density of matter has some particular value.

The only observer to see this model as isotropic would be one situated at its centre. From this privileged position Kant's universe has spherical symmetry on the large scale, but from nowhere else. Similarly, the model could not be described as homogeneous except in terms of a rather artificial definition of homogeneity. Consequently the Kantian model is not in accordance with the cosmological principle as his universe does not present the same aspect to all observers. Perhaps as near as we can come to a regulating principle like the cosmological principle is to say that the observer at the centre, looking to some large distance will see a situation which previously prevailed closer by and will later prevail further off. He might for instance be considering the average density of matter. This does not seem a very satisfying basis for a principle though it could be useful as a guide to surveying the universe.

The Kantian universe has some inherent difficulties and its reliance on unobservable motions must have made little appeal to astronomers but it did gain acceptance for quite a time. Indeed the later researches of

Herschel and those of Lord Rosse gave it strong support but it was irretrievably damaged by the spectroscopic observations that produced such remarkable advances in the next century. Kant's cosmology was a really powerful intellectual achievement and though ultimately discarded, it was the most prophetic that has ever been made. By the 1930's it had been clearly established that the Milky Way is a galaxy and that it is a member of a countless system of galaxies distributed throughout the universe and constituting its principal structural units. Kant was also right in saying that the galaxies are all rotating about their own centres, in fact we do not know of any body in space which does not rotate.

Lambert's model of the universe differs most from Kant's in the introduction of a more complex hierarchy of rotations. In the Milky Way the stars form rotating groups which in turn make up the galaxy. Groups of galaxies spin around centres which again spin round centres and so on. Somewhere in all this is the ultimate centre of centres. Where Kant was satisfied to build his model out of stars and galaxies Lambert went much further in speculation. He postulated for the centres bodies of enormous size, suggesting that the local centre about which the sun moves is a massive body with a diameter at least as big as the orbit of Saturn. This centre he identified with the Orion nebula. The centres of the higher orders must be inconceivably large, giving the whole model a fanciful character sufficient to prevent it from being taken seriously.

Sir William Herschel was an adherent of the theory of uniform distribution for a good many years but later convinced himself out of his own observations that it could not be true. (4) When Herschel undertook the prodigious task of surveying the heavens no one had so far built a telescope adequate for his purpose so he employed his great technical skill to make a reflector of four feet in aperture, then by far the most powerful instrument in existence.

Herschel had to make some estimate of the distance to which his telescope would enable him to see. As no real measures of the distances of stars had then been achieved, Herschel accepted the then current estimate of 3 light-years for the distance of Sirius. On his scale stars at the limit of

---

4. HERSCHEL SIR W., Phil. Trans. Vol. Cl. 1811, P. 269.
   "I must freely confess that by continuing my sweeps of the heavens my opinion of the arrangement of the stars and their magnitudes, and of some other particulars, has undergone a gradual change; and indeed, when the novelty of the subject is considered, we cannot be surprised that many things formerly taken for granted, should on examination prove to be different from what they were generally but incautiously taken to be."

visibility with the naked eye have 1/64th of the brightness of Sirius and, according to the inverse square law are eight times as far away, making the space penetrating power of the eye twenty-four light-years. Comparing the area of the pupil of the eye to that of his 48 inch mirror, since the light-grasp is proportional to area, he concluded that the telescope would enable him to see to a distance of 4,000 light-years.

Having divided the sky into convenient sized regions in which he counted the stars, Herschel proceeded to make 3,400 of these "gauges" as he called them and counted a total of 5,819,000 stars of which over 4,000,000 were in the Milky Way. Using the calculated space penetrating power of the telescope it was possible for him to build up a three dimensional picture of the system of stars. The result was a disc shaped structure about 8,000 light-years in diameter and 1,000 light-years in thickness. Thus by the prodigious labours of a great astronomer, uniformity was finally broken down and the speculative models of the galaxy supported by the best observations that could be made. Though we now know that the galaxy is much bigger than Herschel supposed, it does not detract from his achievement for the errors in his estimates of distance were then unavoidable.

In the course of his surveys Herschel recorded nebulae and globular clusters lying well out of the plane of the galaxy and estimated them to be at distances very large compared with the size of the galaxy, even as far as two million light-years. Such objects he took to be other galaxies out in space and at differing stages of development which naturally lent further support to the theory of Kant.

Herschel did not suppose that all nebulae consist entirely of stars. He thought, as had several others before him, that there exist in the universe patches of what he called "shining fluid" and described as "more fit to produce a star by its condensation, than to depend on the star for its existence." [5]

Another astronomer, Lord Rosse, inspired by the success of Herschel's telescopes and having more than adequate financial resources, built a telescope with a mirror no less than six feet in diameter. By 1850, Rosse [6] was able to announce the discovery of 14 nebulae with spiral structure which did not seem to be parts of the galaxy. At that stage it really looked as if the Kantian model or something very like it was a true representation of the real universe but meanwhile the considerable ad-

5. HERSCHEL, SIR W.. Phil Trans. Vol. LXXXI, 1791, P. 85.
6. ROSSE, Phil. Trans. Vol. CXL 1850, P. 505.

vances in the study of local stars had begun to raise other issues.

Once Bessell had shown that the stars of greatest apparent brightness are not necessarily the nearest and that they can differ greatly in intrinsic brightness, astronomers began to take an interest in the nature of stars. Dynamical studies of double and multiple stars revealed that they differ considerably in mass as well as in brightness. The development of achromatic telescopes which do not introduce false colour effects showed that stars also differ more markedly in colour than appears to the naked eye and this was naturally attributed to differences in temperature. Photometry became a more precise technique and led to the discovery that many stars vary in brightness from time to time, mostly in a definitely periodic manner. It was a time to ask questions about the nature of stars.

About this same time the spectroscope was invented, making it possible to determine the chemical composition of luminously hot materials from the characteristics of the light emitted by them. When it was learned that the dark lines which Fraunhofer had discovered in the spectrum of sunlight reveal the chemical composition of the sun, it was an obvious step to examine the light of other stars. The technical difficulties were finally overcome by Secchi in Italy and Huggins in England who showed that the stars are composed of the same chemical elements as the sun and the earth.

Soon the spectroscopic method became refined enough to deal with the diffuse light of the nebulae. To his great surprise Huggins found that many of them showed, not the continuous spectrum crossed by dark lines which is characteristic of stars, but only the bright lines of glowing gases, mostly hydrogen and nitrogen. [7]

> "The riddle of the nebulae was solved. The answer, which had come to us in the light itself, read: Not an aggregation of stars, but a luminous gas."

With these words Huggins struck a devastating blow at the theory of extragalactic nebulae. Though several of the spiral nebuae, like M31 in Andromeda and M63 in Canes Venatici, displayed a continuous spectrum compatible with a star-like constitution and vast distance, the appearance of a supernova star in M31 was taken as strong evidence against their extra-galactic nature. Apparently another strong point against the Kantian concept was the total failure of much improved instrumental facilities to detect any motion whatever in the nebulae. There was now too much

---

7. HUGGINS, SIR W., Phil. Trans. Vol. CLIV, 1868, P. 540.

evidence althogether against the universe of islands and it was supplanted by the Island Universe.

This new model of the universe was a circumscribed one indeed; except for the lack of an impenetrable boundary it was as restricted as the classical Greek models. There was the galaxy, a finite system of stars and gas clouds, disclike in shape, floating alone in infinite space. Obviously any such structure must have a centre of gravity but by virtue of its rotation centrifugal force would prevent the material universe from collapsing under its own gravitation. Despite its utter isolation the old Newtonian problem of determining rotation relative to absolute space does not arise; the galaxy does not rotate as a solid body, but the stars at greater distances from the centre have longer periods of revolution, according to Kepler's law and so the rotation of the system can be detected from within; thus in principle, the centre of rotation, the centre of gravity indeed, can be located.

The great French mathematician Laplace ([8]) developed a theory of the formation of stars applicable to the island model. In the beginning there had been a huge cloud of gas which contracting under its own gravity, broke into fragments. These fragments, contracting still more, broke up and condensed into stars. The cloud fragments were supposed to be rotating but there is no mystery in this. Given a large mass of material in the process of contracting, it would have to be perfectly symmetrical and the motions of its parts inwards perfectly uniform to prevent it from starting to rotate. The motion of any part a little out of balance with the rest would set the whole spinning and the action would be cumulative. As a rotating body contracts its speed of rotation increases unless it can get rid of some of its angular momentum. In the theory of Laplace this is how the planets were formed. The spinning gas cloud which contracted into the sun threw off some of its material when its speed became so high that centrifugal force overcame gravity and the part thrown off condensed into the planets.

The source of energy which makes the stars radiate light and heat was thought to be the gravitational contraction by which energy of motion was converted into heat. The stars are continually dissipating their store of thermal energy by radiation and thus cooling down, they must after

---

8. LAPLACE, S. P., *Exposition du Système du Monde*, 1796 and further development in the *Mécanique Céleste*. This theory fails to account for the actual distribution of angular momentum in the solar system and is no longer used in its original form. See BOK, B. J., *The Astronomer's* Universe, Melb. Univ. Press, 1958, P. 11, or F. HOYLE, *The Frontiers of Astronomy*, Heinemann, 1955, P. 91.

The 74 inch telescope at Mount Stromlo Observatory, which is largely used for spectrographic work. By means of suitable moving mirrors in the axes the starlight is brought down to a high dispersion spectrograph in a constant temperature room cut out of the solid rock below the lower pier.

The Large Magellanic Cloud, one of the two dwarf galaxies associated with our Milky Way and which may be gravitationally bound satellites. It is about 180,000 light years distant from the sun and occupies about 6° of the celestial hemisphere. This cloud appears to be in rotation and may be developing spiral arms. Mount Stromlo Observatory photograph.

some millions of years go dark. Ultimately all the thermal energy of the material universe must be radiated away to vanish into the depths of space, resulting in the thermodynamic death of the system.

It is small wonder that for the next several generations of men cosmology, faced with a model so limited in every way, was virtually abandoned. Quite apart from the restrictions imposed by the discoveries of physical science, the latter half of the 19th century was the heyday of mechanistic determinism. This was a mode of thought, amounting to a creed, in which the possibility of labelling something as philosophical, let alone metaphysical, was sufficient to damn it utterly in the mind of almost everyone with any pretensions to being a scientist. The immediate successors of Newton felt that the world was the work of a supermathematician but, for 19th century science, it was the work of a supermechanic; everything could be reduced to mechanical principles and all the important things had now been discovered, leaving only the details to be filled in by the scientists of the future.

While physical scientists felt secure in their achievements to hold forth dogmatically about the future, trouble was brewing in several quarters. The geologists were not at all satisfied with the life time of the sun and, by implication, of the earth, which contemporary physics would allow. If the sun were simply cooling down it could be of no great age. An attempt was made by a French scientist to extrapolate the observed cooling rate of hot metallic spheres observed here to the sun, but no few million years calculated from the cooling of a hypothetical iron ball the best part of a million across would suffice. An attempt to give the sun a source of energy from the infall of meteorites drawn in by gravity was also unsuccessful, for it was calculated that this would demand an influx of matter at the rate of 9.4×10^{13} g/sec., which, in an astronomically short time, would so increase the mass of the sun that the dynamics of the solar system would be upset. Lord Kelvin showed that this would involve the shortening of the year by a day in every 50 of our present years.

The only possible source of energy for the sun then seemed to be that of gravitational contraction. A contraction of about 100 yards or metres a year would be enough to maintain the observed solar emission of light and heat, but this can be shown to make the sun only some 20 million years old. [9] Proposed by Helmholtz in 1854 this seemed to be the only

---

9. A number of contemporary opinions are given by CLERKE, *A Popular History of Astronomy in the Nineteenth Century*, 1885, P. 350 ff. On the most favoured estimate, the sun would have a total luminous life of about 30 million years.

tenable answer until early in this century when Rutherford could offer an extension in terms of the energy released by natural radioactivity, later to be followed up by the present theory of nuclear fusion. Geologists are now content with a solar history of a few thousand million years and a future perhaps three times as long.

Biologists were also unhappy about the short time scale because the then recently published Darwinian theory of evolution demanded much more than 20 million years. On quite different grounds many biologists were at odds with 19th century physics: they were unwilling to accept a purely mechanical basis for life. There seemed to be something more involved, a vital force unknown to physics. While vitalism has long since had its day, it was then a power to be reckoned with. What remains of it today is deeply submerged in the general question of teleology. It is very easy indeed to become involved in an argument as to whether or not evolution is in some sense purposeful. Of course teleological argument does not stop there; the whole structure and history of the universe can be brought into the discussion. There is a strong tendency in teleology which has been apparent since at least the times of St. Thomas Aquinas to add up a series of individually not quite conclusive arguments into a whole which is apparently stronger than the sum of its parts. The approach often takes the form of an intuitive assessment of probabilities: many features of the universe can be pointed out which, taken invdividually may be attributable to pure chance, but, taken as a whole, the probability of them all occurring by chance is considered too overwhelmingly small to be accepted. Such an argument has often been used about the occurrence of life in the universe. Nowadays we are more inclined to view our part of the universe as typical and since life *has* occurred here, its probability is necessarily greater than zero, so we can expect it to occur elsewhere also. On the basis that our sun is a fairly typical star we can reasonably suppose that life resembling ours could be found in other parts of our own galaxy.

Even though the island universe might well contain inhabited planets other than our own, it has little appeal to cosmological thinkers; it is depressingly small in extent and short-lived in time to be the sum total of things. Yet the acceptance of such a model might have been forced upon us for much longer had not physics itself contained the seed of disruption. The source of the upheaval that was soon to come lay well within the domain of physics, in the phenomena of light.

And lo! The Hunter of the East has caught
The Sultan's Turret in a Noose of Light.

CHAPTER 8

LIGHT AND THE ETHER

Light has always been our principal source of information about distant bodies; indeed for objects at astronomically significant distances it is virtually the only source. We are no longer limited to the visible spectrum of light as the range of observations has now been extended into the infra-red and radio regions on the one side and into the ultra-violet and X-ray radiations on the other. It is common knowledge that these radiations are all of the same nature as light, even though they are observed with quite different kinds of apparatus. The nature of light is a proper subject of interest for astronomers and cosmologists alike but detailed investigations into it must be left to the more specialised physicists. Up to a point we may consider ourselves as mere users of light and accept whatever theory of it seems the best to our colleagues but sooner or later that theory must prove compatible with observations made in other branches of science, more particularly, our own.

The velocity of light is a very important quantity for us because of its intimate relationship with distance. We have already seen that it can be used to express distance very conveniently in terms of time, but, we can go further than that, and use it, if we wish, and as did E. A. Milne, to define distance in terms of time. ([1]) Since the development of radar in the second world war there is nothing very novel about the actual measurement of distance in terms of time. The principle is rather simple; a flash of light or a pulse of radio waves is transmitted to the distant object and some of it is reflected back. The interval of time between sending out the pulse and receiving the reflection is measured and knowing the velocity of light, half that time is the measure of the distance. The secret of success for small distances lies in being able to measure extremely short intervals while for large distances we must be able to send out exceedingly powerful pulses and detect very weak reflections. But the concept of distance here is still the classical one of a measuring stick,

---

1. For a convenient summary of Milne's proposition see BONDI, Loc. Cit. P. 27.

a rigid ruler. When Milne proposed the definition of distance in terms of time, he wanted to get rid of the ruler altogether. To do this it is necessary to forget about the velocity of light as so many miles per second, because that presupposes the ruler: it must be taken simply as a natural constant and for reasons of mathematical convenience, it is often assigned the value of unity. On this basis the quantity ct, where c is the symbol for the velocity of light and t is the time of transit, specifies a distance. [2]

At the time when Milne put forward this concept of distance in the 1930's, most cosmologists objected to it on the grounds that while it might be conceptually possible to determine distance in this way, there was no possibility of actually doing it. None had then found a way to measure the minute intervals involved for short distances and for long distances it was obvious that, even if reflections could occur, the waiting time would be much longer than the lifetime of any human observer. However, since then engineers have solved the short time problem to the extent that radar devices are now in regular use as surveying instruments of great precision. Medium distances have also proved manageable; already the distances of the moon, the sun and the inner planets have been determined by radar to greater accuracies than ever before. Of course the long time scale remains an insuperable difficulty for cosmic distances, even if the technical difficulties of transmission and reception could be overcome.

Even though most cosmologists do not choose to adopt it, a concept such as Milne's to which no logical exception can be taken and which is capable of application on the scale of the astronomical unit, cannot be condemned out of hand. If nothing else, it is salutary for us to reflect that time and space are not necessarily the completely disparate entities that we commonly assume them to be and that classical scientists firmly believed them to be.

The velocity of light was first successfully determined by Olaus Roemer in 1675. [3] He found that the eclipses of the moons of Jupiter were delayed from their calculated times according to the position of the earth in its orbit. The difference when the earth was more distant from Jupiter by the diameter of its orbit amounted to about 22 minutes. This Roemer correctly attributed to the time taken by light to traverse that distance and so he could calculate the velocity in terms of the then current value of

---

2. This notation is widely used in cosmology where time and space have to be brought on to the same mathematical footing.
3. Reprinted in SHAPLEY & HOWARTH, Loc. Cit. P. 70.

the astronomical unit. Both this unit and the time were rather inaccurately known then, but we can take more modern figures, 186 million miles, 300 million km, for the diameter of the orbit and 1,000 seconds for the time, giving 186,000 miles, or 300,000 km per second for the velocity of light.

Bradley, [4] also found a value for the velocity of light by an astronomical method. His discovery of the aberration of light leads to a value of the velocity, since the annual change in position of stars is determined by the ratio of the velocity of light to the velocity of the earth, which is known. Though Bradley thought that his elimination of an annual displacement which had previously been taken for a parallax might be used as evidence against the Copernican theory, the observed motion of aberration is consistent with both the Copernican and Ptolemaic theories. By Bradley's time astronomers generally had adopted the Copernican theory though churchmen were still opposed to it on principle.

Although the astronomical methods showed that the velocity of light is enormous by ordinary standards, ingenious experiments were devised to measure it under laboratory conditions. First Fizeau, using a rapidly rotating toothed wheel and then Foucault, [5] using a rotating mirror, were able to chop a beam of light into extremely short flashes. These after reflection from a distant mirror returned to the rotating mechanism which meanwhile had moved enough to block or displace them. As the distance travelled and the rate of movement of the wheel or mirror were known the velocity of light could be found. The values so found were in reasonable agreement with those determined astronomically and are in still better agreement with those determined more recently with improved technical facilities.

It would be interesting to know whether the velocity of light remains constant or whether it changes with the passage of time. If there has been any change during the period that we have had reliable values it is smaller than the experimental errors of measurement. For cosmological purposes we can be content with the rounded-off value of 186,000 miles or 300 million metres per second for the velocity of light in vacuo. It is slightly

4. Ibid, P. 103.
5. Detailed accounts of these experiments can be found in most textbooks of Physics. For these and other methods see also JENKINS & WHITE, *Fundamentals of Optics*. McGraw-Hill, 1957, Chap. 19.

less in air and considerably less in denser media ([6]) but light of interest to cosmologists will have taken only an utterly insignificant part of its passage through anything but vacuum.

For a span of two thousand years after Empedocles ([7]) the corpuscular theory of light prevailed. ([8]) According to this theory, light consists of minute material particles emitted by luminous bodies and travels in straight lines in free space. The phenomena of vision arise from the impact of the corpuscles on the retina of the eye, where they set up vibrations detected by the fibres of the optic nerve. Similarly light causes heating of materials which absorb it by the agitation of their constituent molecules. In terms of the corpuscular theory the laws of reflection and refraction of light and hence the whole study of geometrical optics can readily be developed.

To account successfully for other phenomena of light such as come into the study of physical optics where the nature of light is of deeper concern, the light corpuscles have to be endowed with rather remarkable properties. As material bodies which in principle can be neither created nor destroyed they must exist in unlimited numbers in all matter or alternatively there must be some mechanism of transmutation between light and normal matter. A sufficiently hot body will continue to emit light for so long as it can remain hot and in the case of the stars, this can be a very long time indeed. It follows that the corpuscles must be vanishingly small in mass and in size to prevent the effects of those properties from being detectable in emitting and absorbing bodies. If the mass of the corpuscles were appreciable then their integrated momentum in a strong beam travelling at 186,000 miles per second would be not

---

6. The velocity of light in a dense medium is related to the refractive index of the material thus: $n = \dfrac{c}{v}$ where v is the velocity in the medium and n is its refractive index. However it is necessary to distinguish between wave or phase velocity and group velocity. For water waves it is easily seen that the wave velocity is greater than the group velocity by dropping a stone into smooth water and observing that the ripples rise at the rear of the group and move forwards through it. For light, the direct measurements of the Fizeau/Foucault type measure the group velocity as the light is chopped into pulses containing a large number of waves. In a material medium where the group velocity is lower than c, its value depends on the wavelength of the light within the medium according to Rayleigh's equation $u = v - \dfrac{\lambda dv}{d\lambda}$, where u and v are the group and phase velocities respectively. Thus in a dense medium the group velocity of light differs for the various colours by a small factor, red light travelling faster than blue. In vacuo u is equal to c for all wavelengths.

7. Empedocles, a Greek philosopher of the 5th century B.C. seems to have been the first to formulate a corpuscular theory.

8. The final blow to the old corpuscular theory was delivered by Foucault in 1850.

merely detectable but positively destructive in its effects. We also know that two beams of light can intersect without affecting each other but it would be reasonable to expect two concentrated streams of material particles, like jets of water or air, to disturb each other and cause some scattering of the light.

The colours of light present another problem soluble only by the hypothesis that for each colour there exists a specific variety of the basic type of corpuscle. In view of their evanescent size and mass it seems difficult to conceive that one of these properties could be modified but something of the sort is necessary because astronomical observation shows that the velocity of light is the same for all colours. ([9]) Newton adopted the hypothesis that the corpuscles differ in size, the largest exciting the sensation of red light and the smallest, violet.

About 1810, Malus ([10]) established the polarisation of light, working on the phenomena of double refraction which had concerned Huygens so long before. It appeared that by passage through certain crystalline substances in the appropriate plane of their crystal structure, the light corpuscles became so aligned that a beam of light would readily pass through a second such crystal in the same orientation but could not pass when the plane of transmission was turned 90° about an axis parallel to the direction of the beam. To explain this result Malus had to postulate a polarity in the corpuscles which might be supposed to originate in the geometry of their shape but which he seems to have thought to arise from some property analogous to magnetic polarity.

Thus we see that as various optical phenomena came up for explanation new and strange properties had to be attributed to the corpuscles with the result that they became so artificial in character that they were conceivable only by an act of faith. Indeed, the only really attractive feature of the corpuscles was that they provided such a simple explanation for the propagation of light in empty space.

Newton was a supporter of the corpuscular theory but it is obvious to anyone who reads his "Opticks" ([11]) that he was far from satisfied with it as a comprehensive theory. Newton took up an attitude more com-

---

9. Consider a typical variable star of the eclipsing binary type, for which the changes in brightness are caused by a dark companion periodically coming between it and us. No such star is closer than a few light-years and, if light of different colours had different velocities *in vacuo*, we would see changes of colour as well as of brightness. In a travelling time of some years even very small differences of velocity would become manifest.

10. STRONG. J., *Concepts of Classical Optics*, Freeman. 1958. P. 110

11. NEWTON, SIR I., *Opticks*, Dover reprint S 205. 1952.

patible with those of the present day than most of his nearer successors would have countenanced: he simply adopted those features of the rival theory that suited his purpose and were reasonably reconcilable with the corpuscular theory. In his communication of 1675 to the Royal Society[12] we find: —

> "For assuming the rays of light to be small bodies emitted every way from shining substances, those, when they impinge on any refracting or reflecting superficies, must as necessarily excite vibrations in the aether as stones do in water when thrown into it."

and in another sentence he says: —

> "... because I see not how the colours of their transparent plates or skins, can be handsomely explained without having recourse to aetherial pulses."

To appreciate how Newton could do this and to understand the basis of the wave theory of light we must now look at the concept of the ether.

When we speak of the ether it is usually in reference to the luminiferous ether first clearly proposed by Huygens, the real father of the wave theory, [13] as the vehicle for its propagation. Though it was by far the most important and long-lived many other ethers have been invented to explain among other things, electric, magnetic and chemical effects. The ethers were all conceived as material in nature however subtle and tenuous they might be and many of them were supposed to fill all space. In the context of Newtonian dynamics such a medium, all-pervading, uniform and devoid of large scale motions, must coincide to a great extent with absolute space. As physical science developed, interest in absolute space diminished almost to mere lip service, there being no need for reference to it to appear in the vast majority of the formulations of scientific laws. On the other hand the luminiferous ether grew in importance and incidentally came so near to fulfilling every purpose for which the concept of absolute space had been devised that it virtually became absolute space in material embodiment. In a very real sense it can be called "the ether of space." Scientists tended to forget the hypothetical character of the ether and to take it for a physical reality; thus Clerk Maxwell said: —

> "Whatever difficulties we may have in forming a consistent idea of the constitution of the ether, there can be no doubt that the

12. Reprinted in BREWSTER'S *Memoirs of Sir Isaac Newton*, Constable 1955, P. 391.
13. HUYGENS, C., *Traité de la Lumière*, 1690, Translation by S.P. Thompson, Dover reprint S 179.

interplanetary and interstellar spaces are not empty, but are occupied by a material substance or body, which is certainly the
largest, and probably the most uniform body of which we have
any knowledge."

Descartes had to postulate an ether on philosophical grounds. In the
Cartesian scheme of things, extension is the essential and exclusive property of material bodies: nothing but a material body can have size; and
distance consists in a succession of contiguous bodies. Yet it is obvious
that in the world as we know it bodies do exist at great distances apart
in apparently empty space. The ether solved this problem by abolishing
empty space and providing a material medium in which distance could
materialise.

The idea of action at a distance whereby one body can influence
another across a span of empty space is repugnant to the human mode
of thought. It was thus natural for thinkers to propose ethers to account
for phenomena seemingly operating across empty space. Gravity was an
obvious case and Newton postulated an ether to account for its action at
a distance. The strength of his feeling on this point is revealed in one of
the letters to Bentley. ([14])

"That gravity should be innate, inherent, and essential to matter,
so that one body may act upon another at a distance through a
vacuum, without the mediation of anything else, by and through
which their action and force may be conveyed from one to
another, is to me so great an absurdity that I believe no man
who has in philosophical matters a competent faculty of thinking
can ever fall into it."

Huygens founded his wave theory on an analogy with sound which was
known to be a periodic disturbance in the air. Sound waves are of the
longitudinal kind consisting in a succession of alternating regions of compression and rarefaction moving through the air with a velocity determined
by its properties. ([15]) Huygens supposed that light waves are also longitudinal and this meant that his ether had to be a fluid medium composed
of particles. Sharing the general aversion to action at a distance Huygens
followed Descartes in making the ether particles very minute and elastic

14. Reprinted in MUNITZ, Loc. Cit. P. 217.
15. The velocity of sound in air depends on its density and bulk modulus, defined
in this adiabatic case as $B = \dfrac{\text{Change of pressure}}{\text{Change of volume per unit volume}}$

Where ϱ is the density we have $v = \dfrac{B}{\varrho}$

with successive orders of smaller particles filling the interstices between the larger so that in the end there was no empty space whatever.

The longitudinal wave theory could offer explanations of reflection, refraction and double refraction and the question why light does not diffuse around corners like sound was answered by Huygens by reference to the extremely small wavelength. The analogy with sound applies here also, for the shorter wavelength of the sound the more directional is its propagation.

Newton ([16]) rejected the longitudinal wave theory and it appears that the etherial vibrations he had in mind were the transverse kind, for he referred to them as having "sides." Both Newton and Huygens knew that, as early as 1665, Hooke had suggested the possibility that light might be a transverse wave phenomenon but Hooke was one of those unfortunate people who have ideas but lack the application to build theories upon them. Consequently, since Huygens kept to the longitudinal theory which the great authority of Newton caused others to neglect and Newton himself was interested only in ether waves as a purely local phenomenon where light interacted with matter, it was a long time before anyone took up a wave theory of light again.

Finally at the beginning of the 19th century Young became interested in the possibilities of a wave theory and, like Huygens, began with longitudinal waves. Meanwhile Fresnel had also taken up the subject but in terms of transverse waves to which he succeeded in converting Young. The new theory was not readily accepted by the leading authorities of the time, as witness what Laplace said in a letter to Young in 1817: ([17])

> "Je conviens que les nouveaux phénomènes de la lumière sont jusqu'à présent très difficiles à expliquer; mais en les étudiant avec un grand soin, pour découvrir les lois dont ils dépendent, on parviendra peut-être un jour à reconnaître dans les molécules lumineuses des propriétés nouvelles qui donneront une explication mathématique de ces phénomènes."

There were two major points at issue between the corpuscular and wave theories. One was the interference of light which the wave theory could readily explain but the other could not. By now the eclectic outlook of Newton had been abandoned and the corpuscular theory was strictly corpuscular, so local wave effects could not be called upon to help. Interference is the phenomenon observed when two related beams of light

---

16. NEWTON, Loc. Cit. P. 362 ff.
17. From Young's "Works" Ed. Peacock. Quoted by Tait, P. G., Encyc. Brit. 9th Ed. sub voce "Light".

are brought together and it is found that in the region where they co-exist the two beams alternately reinforce and extinguish each other. The experiment is commonly performed by passing light of a single colour through a narrow slit and then, at a little distance, through two similar slits side by side and close together in an opaque sheet so that the light falls on to a white surface set some additional distance away. Then, instead of uniform illumination, a series of bright and dark bands appear on the screen. This is readily explained by the wave theory; the two sets of waves have a constant phase relation since they come from the same source and where they strike the screen there are places where the crests of the two sets of waves coincide producing a bright band and others where the crests of one train coincide with the troughs of the other so that cancellation leaves dark bands there.

The other point was that the corpuscular theory explained refraction by a greater velocity of light in a dense medium than *in vacuo* while the wave theory makes it less. For water and air the corpuscular ratio is 4/3 whereas on the wave theory it is 3/4. The method by which Foucault had found the velocity of light in air could be used to find it in water by letting the light travel through a long tube filled with water and when this was done, the prediction of the wave theory was found to be correct: the velocity of light in water is only three quarters of the velocity in air. The wave theory was generally accepted after these conclusive demonstrations but several problems still remained, arising out of the nature of transverse waves and centring around the phenomena of polarisation.

To become a medium suitable for the propagation of transverse waves the character of the ether had to be substantially altered; instead of an elastic fluid it had to be an elastic solid. It is customary to use water waves as an analogy for ether waves and the analogy works very well within limits. Water waves are transverse but they are confined to the surface. In general, transverse waves in a fluid can be propagated only along its interface with another fluid of different density and in the presence of a force tending to restore the level of the interface. In the usual case the interface is between water and air and the restoring force is gravity; for waves of very small amplitude it may be surface tension. Although a transverse wave can be generated wholly within a fluid it cannot be propagated to a distance as it dies away in much less than its own wavelength. But in an elastic solid medium such waves can be propagated to great distances if the frictional and other losses of energy be small enough.

As with surface waves on water, the vibration of transverse ether waves is at right angles to the direction of propagation and visualising a train of waves approaching along the line of sight with the oscillation in the vertical plane, it is possible to gain a notion of what Newton meant by saying that they have sides. The power of visualisation has been of enormous value to mankind in technology and in many fields of science but for the kind of concept which concerns us now it is inadequate and consequently misleading. The behaviours of plane polarised light is best explained by supposing that the vibrations occur in the plane of polarisation but reflection can be better understood on the hypothesis that the vibrations are at right angles to that plane. Later the electromagnetic theory developed by Clerk Maxwell (18) involved vibrations in both planes, one electric and the other magnetic, not merely related but indissolubly bound together by the continual sharing and interchange between them of the energy of the wave motion. Complex disturbances of this kind, Maxwell showed, must arise from absolute motion through the ether of electric charges. It had not previously been necessary to assume that electric and luminiferous ethers were one and the same but when he found that these new disturbances would have the same velocity as light, there was only one conclusion to be drawn. In due course the predicted waves were found by experiment, largely through the work of Hertz who was the first to generate and detect radio waves. Thus the Maxwellian theory united the whole gamut of radiations from the longest radio waves, through radiant heat and light, on to X and gamma rays. They are all electromagnetic waves differing only in wavelength.

In free space all these waves are propagated at the same velocity and in precisely the same way but in their interactions with matter the differences in wavelength, or frequency which we now commonly find it more convenient to think about, produce vastly different phenomena. Wavelength, the distance from crest to crest and frequency, the number of waves passing a given point in unit time, are reciprocally related by the velocity. Putting n for frequency and λ for wavelength, we have $c = \lambda n$. When we consider that the waves are a disturbance of the medium it is not surprising to find that higher frequency, thus shorter wavelength, means greater energy in the waves.

The electromagnetic transverse wave theory of light provided an astonishingly successful explanation for all the phenomena of light known up to the end of the 19th century when the photo-electric behaviour of the newly discovered electrons proved too much for the classical Maxwell-

18. STRONG, J., Loc. Cit. P. 46 ff.

ian version and necessitated further modifications. It is hardly to be wondered that despite a few minor difficulties the nearly infallible wave theory should have lent so much strength to the sense of reality which the scientists of those days had for the ether.

Now we must consider the properties which the ether must have as the vehicle for the propagation of light, making occasional circumspect use of the analogy with water waves. Firstly since the absorption hypothesis of Cheseaux and Olbers no longer applies, light must be able to travel indefinitely far through the ether without loss of energy; this means that it must be completely frictionless and free from any other dissipative losses. Because of its finite velocity the energy of light in transit must exist for some time in the ether and at any moment half of this energy is in the form of kinetic energy of motion in the medium and half as potential energy in the state of strain of the medium. In water waves the potential energy component is found where the water is raised above and depressed below the undisturbed level while the kinetic energy resides in the forward motion of the humps and hollows which is an oscillatory motion of the water, not a general motion of translation. The water only acquires a bulk forward motion when the waves come to the shore where the water is shallow in comparison to the amplitude of the waves. On a large enough surface of deep water otherwise undisturbed the waves will die away because friction dissipates their kinetic energy.

The ether has to behave like an elastic solid to provide the restoration to normal which gravity does for water. We may perhaps, think very naïvely of the ether as having a sort of rubbery consistency, making it always want to return to its original state. As an elastic solid the ether will also have rigidity. The magnitudes of the elasticity and rigidity of the ether had to be really different from those of ordinary matter, compared with materials like steel or plastics. (19) The differences are in the order of a hundred million times. These are obviously inferred magnitudes as the only measurable quantity involved is the velocity of light. It amounts to saying that since we are considering light as a mechanical

---

19. The following values were found by Clerk Maxwell on the basis that the density of radiant energy close to the sun is 1.886 erg./cm^3, and that the amplitude of the waves is not more than 1/100 wavelength.

Density of ether 5.36 $\times$ 10^{-19} g/cm^3
Elasticity (shear stress) 30.17 dyne/cm^2 ($\varrho c^2 \cdot 2\pi/100$)
Rigidity 482 dyne/cm^2 (ϱc^2)

The coefficients of rigidity and elasticity of metals and plastics are a few times 10^{11} dynes/cm^2

wave disturbance propagated in the medium it must have these extra-ordinary properties to account for the enormous velocity of light compared with the velocities of disturbances in ordinary matter. We must not lose sight of the basic idea that nineteenth century science wanted a mechanical explanation for the phenomena of electricity, magnetism and light.

A strange and apparently contradictory property of the ether is that it must be incompressible. Mathematically the wave theory admits the possibility of disturbances at right angles to the wave surface, that is, longitudinal waves could occur. As there are no phenomena whatever to be observed that could arise from this, it was taken that they cannot occur which is only the case if the ether is incompressible.

To provide for the storage of potential energy the ether should have a certain density which together with the elasticity, will determine the velocity of propagation through it. The density comes out as 5×10^{-19} g/cm^3. This is small indeed compared with the density of matter to which we are accustomed on the earth, in the order of grammes per cubic centimetre but on the cosmic scale it is alarmingly large. The average density of matter in the observable universe can be estimated by treating the presently known quantity of matter in large volumes of space as if uniformly spread out, from which we obtain the figure 5×10^{-29} g/cm^3. It looks as if the mass of the ether would outweigh that of normal matter by a factor of one hundred million! Here appears quite a paradox, for the island universe, floating in an infinite ocean of ether, ought to be beset by the same gravitational problem that afflicted the original Newtonian model.

The properties which we must assign to the ether for the propagation of light are all very well providing that we do not ask questions about what the consequences may be in other fields; as we have just seen, the density of the ether is a case in point. Obviously all the celestial bodies, satellites, planets and stars are able to swim freely in the endless ocean of ether, their motions unimpeded by its small but finite density and its rigidity. It is necessary to suppose that either this solid material can penetrate through every kind of matter or that matter can move through it without resistance. Any resistance to motion of translation through the ether must involve dissipation of the kinetic energy of a celestial body and would gradually slow down its motion. Energy so abstracted is in the long run radiated away into space and thus ultimately becomes lost from the material universe. To make a limited analogy disregarding friction, the water waves it sets up carry away energy from a moving boat and without a continuing source of energy, the motor, it would slow down.

Already serious mechanical difficulties have appeared but the situation is far worse when mechanics and the electrodynamics of light have to be taken into account together. On the large scale the ether is motionless but the earth is in motion relative to it so there ought to be an "ether wind." If such a phenomenon is to be detectable experimentally it will have to be in terms of the velocity of light for lack of any other factor, since the necessary characteristics of the ether have made it otherwise inaccessible. Sound and light have in common the fact that their velocities through their respective media are determined solely by the properties of those media. Both sound and light exhibit what is known as the Doppler effect. If the source of a steady sound tone and the observer are approaching one another, the observed pitch of the tone is higher than if there were no relative motion and the opposite effect is found when the motion is one of recession. The cause of the phenomenon is simple enough: a motion of approach results in the observer receiving more waves per second in the first case and fewer in the latter, in other words, the pitch or frequency is altered by the motion between the source and the observer and it does not matter which may be in motion and which at rest. For light, as has been mentioned in an earlier chapter, the Doppler effect results in a change in colour, the optical equivalent of sound pitch, and is used to determine relative velocities in the line of sight. In a later chapter we shall be concerned with the magnitude of the Doppler effect in light.

In an experiment with sound it is quite easy to demonstrate that if the air between the source and the observer is moving towards him, the velocity of sound relative to him will be higher than if it were not moving, because the first waves reaching him will have covered the same distance in less time and *vice versa*. The same applies if the observer or the source is in motion and, in these days of supersonic aircraft, it is well known that an observer can exceed the speed of sound so that relative to him, it can become zero or even a negative quantity. Let us be quite clear about the measurement of the velocity of sound: the observer really measures the velocity of sound relative to himself, counting his apparatus as an extension of his sense organs. When he is stationary relative to the air the measured velocity is the velocity of sound through the air. The indirect method of finding by other means the density and bulk modulus of the air and then calculating the velocity also gives the speed of sound

---

20. BONDI, Loc. Cit. P. 45 f.

through the air. In all this the Doppler effect is purely incidental and arises from the constancy of the velocity of the waves in the medium.

Assuming the validity of Newtonian dynamics, precisely analogous effects should be found for the velocity of light. The situation is basically the same only the experimental conditions are more difficult because the velocity of light is enormous compared with that of any moving observer and, because owing to the motion of the earth, there is no possibility of being an observer at rest relative to the ether. Further the indirect method cannot be used since the properties of the ether are derived from the observed velocity of light. The earth has several motions relative to absolute space, that is relative to the ether, so in principle it should be found that the velocity of light coming from different directions differs and thus reveals the ether wind. Obviously the most convenient known velocity for an observer is that of the earth in its orbit round the sun, which at 18 miles per second, is only one ten thousandth part of the velocity of light so the experimental technique must be very refined if any result is to be obtained.

A very beautiful experiment was devised by Michelson and Morley ([21]) after many others had failed, an experiment so delicate that it would reveal a drift through the ether no greater than a tenth of the earth's velocity. Mounted on a massive block of stone was a lamp delivering a beam of light that could be split into two components by a semi-transparent mirror and made to travel over paths at right angles to each other before being reflected to a common observing point by other mirrors. By making the two light paths as nearly equal as possible, the reunion of the light caused interference patterns to appear at the observing point because of the phase differences of the waves travelling over paths differing by a very small amount; the difference must be a whole number of wavelengths plus a fraction of a wavelength. For green light the wavelength is about half a micron, which unit is the thousandth part of a millimetre, Fig. 5.

The whole apparatus was floated on mercury so that it could readily be turned around to permit one light path to be set in the direction of the earth's motion with the other one at right angles to it and then to interchange them. Any difference in the velocity of light along the two paths would be revealed by a change in the interference pattern. The experiment was repeated many times and with progressively improved apparatus, the utmost precautions being taken to eliminate every kind

---

21. A typical account is in JENKINS AND WHITE, Loc. Cit., P. 399 ff.

of error. The results of the experiment were always the same: no ether wind was ever observed. (22)

There could be no doubt about the sensitivity and reliability of the Michelson-Morley experiment and, from the time when its results were finally made known in 1887, consternation prevailed in scientific circles. It is easy enough for us with the wisdom that comes after the event and having in mind the totally contradictory properties of the ether to think that the Michelson-Morley experiment simply proved that this utterly

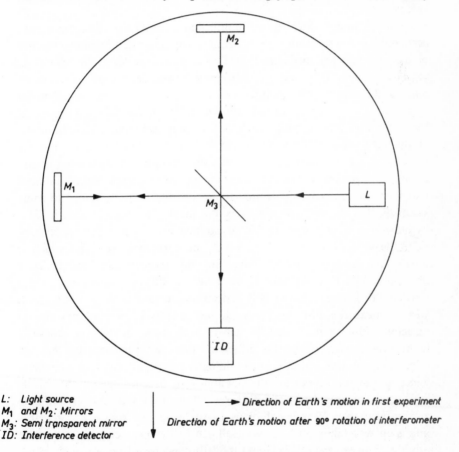

L:　Light source
M_1 and M_2: Mirrors
M_3: Semi transparent mirror
ID: Interference detector

⟶ Direction of Earth's motion in first experiment

Direction of Earth's motion after 90° rotation of interferometer

Michelson's interferometer on a rotatable stone slab floating in mercury

Fig. 5　The Michelson and Morley Experiment.

22.　When repeating the experiment after an interval of six months the earth is at a diametrically opposite position in its orbit and its orbital velocity through the ether is reversed.

improbable medium does not exist, forgetting conveniently for the moment the question of action at a distance. Strictly we should say that it proves there is no specially privileged frame of reference for absolute motion, that is, no absolute space; and by implication, no absolute time. After all there is nothing very absolute about scientific theories; in general, conclusions are reached in practice on the balance of probabilities. The success of the ether wave theory in optics was too great altogether for it to be quickly discarded; perhaps there were other possibilities of saving the theory.

Although matter could move unimpeded through the ether there must certainly be some interaction between them and experimental evidence of this was already available. [23] The crucial experiment by which the velocity of light in water was determined could easily be modified by making the water flow rapidly along the tube. When this was done it appeared that the observed velocity was different for different speeds of the water, not fully in proportion, but at least there was a definite effect of the kind sought. While we might be more inclined to ask what the velocity of light in water has to do with the ether, the interpretation placed upon the experiment was that the moving water carries along the ether that is in and immediately around it. In a similar way it seemed to some scientists that the earth would carry the local ether around with it. This hypothesis is clearly unworkable; if nothing else, it is necessary to consider the velocity of the sun, in which the earth partakes, and which is larger than its own velocity about the sun; there would have to be a discernible effect on the aberration of light in different directions.

But 19th century physics still had an answer and a very subtle one it was: to make the ether unobservable, not merely in practice but also in principle. Philisophers, who in general are more inclined to consider themselves strictly bound by principles than are pragmatically minded scientists, may well object to maintaining the physical reality of something which is put totally outside the possibility of observation by any means whatever. This involves a hypothesis which is at the one time unverifiable and unfalsifiable. However dubious its philosophical status, this remarkable step was taken and, as we shall see in the next chapter, it formed a bridge between the classical and modern concepts of physics with which cosmologists are particularly concerned.

---

23. Ibid P. 396. The experiment was performed in Fizeau in 1859.

For "Is" and "Is not" though with Rule and Line
And Up-and-Down without I could define.

CHAPTER 9

LIGHT AND RELATIVITY

The means by which the ether wind could be made unobservable was proposed independently by Fitzgerald and by Lorentz. [1] The idea was that a moving body may be supposed to contract along the direction of its motion by an amount depending only on its velocity relative to the ether. Thus the apparatus of the Michelson-Morley experiment contracts in the direction of the earth's motion just sufficiently for the shortened light path in that direction to compensate for the expected difference in the measured velocity of light. That matter might be so distorted by motion through the ether is hardly more astonishing than the notion that it would be totally unaffected by passage through a material medium of finite density. For velocities such as the earth possesses, the contraction is extremely small and Lorentz was able to give a theoretical explanation of it consistent with the electronic theory of matter then in vogue.

Obviously the Lorentz contraction cannot be detected by any means of measurement, however refined, because any measuring device laid alongside the Michelson-Morley apparatus will contract by exactly the same amount. It can also be shown that any other indirect means of measurement must be affected in the same way. However, it is not difficult to calculate the magnitude of the contraction for a given velocity. The formula which has since come to be known as the Lorentz transformation, has the form $\sqrt{(1 - \frac{v^2}{c^2})}$, where v is the velocity of the moving body and c, as usual, is the velocity of light. Thus if a metre rod be set in motion lengthways, its length will contract so that

$$l = l_0 \sqrt{(1 - \frac{v^2}{c^2})}$$

---

1. LORENTZ, H.A., *Michelson's Interference Experiment*, in Dover reprints "The Principle of Relativity" S81. P. 3 ff.

and for the velocity of the earth, the contraction amounts to 1 part in 200 million. ([2])

Now on the principles of Newtonian dynamics, there is at first sight, nothing very special about the velocity of light. No reasons appear why a force applied for a long time should not accelerate a body to as high a velocity as we please, much higher for instance, than the velocity of light. We can imagine a remarkable consequence which might arise from so fast a motion through the ether, analogous to the well known sonic booms produced by aircraft flying faster than the speed of sound. It is conceivable that shock waves might be set up in the ether, moving at the speed of light and in principle they should be detectable. ([3]) Perhaps their existence could be inferred in another way; as the shock waves would carry away energy from the moving body; its velocity would fall back again to the speed of light unless we could arrange to supply it continually with energy to make up the loss. If this were the case bodies moving faster than light would apparently not obey Newton's first law of motion.

While this piece of speculation is not meant to be taken as something which could really happen, it will have served a useful purpose if, through it, we perceive more clearly that there is nothing sacrosanct or inviolable about Newtonian dynamics; they are man-made rules, not ultimate truths.

For bodies travelling close to the velocity of light, the Lorentz contraction has a strange consequence. The factor $\dfrac{v^2}{c^2}$ approaches unity very rapidly as v comes close to c. At a velocity of $\frac{1}{2}c$ our metre rod would be only 0.866 of its rest length, at 0.9 it would be less than half and at 0.99 c a little under a seventh of its rest length. When $v = c$ it would have no length at all, which is a physically awkward situation.

When we are faced with consequences of this kind it is time to look more closely at what we are doing, for it probably means that we have imposed some hidden manmade conditions on the phenomena of nature.

2. The Lorentz transformation appears in the mathematics of the Michelson-Morley experiment as the factor determining the distance the apparatus will have moved during the time taken by the light to go and return along the path at right angles to the path in the direction of the earth's motion. For $v = 0$, this distance is zero and a contraction of the path in the direction of motion by this factor will prevent shifting of the interference fringes for any value of v. See also EINSTEIN A., *Relativity*, 1916, Bonanza paperback 1961, Chap. XI.

3. In a material medium where the velocity of light is considerably lower than C, it is possible to make material particles move faster than light and just such disturbances are set up, visible as blue light. These disturbances were first seen in nuclear reactors and are known as Cherenkov radiation after their discoverer.

Obviously the Lorentz contraction works well for small values of v but we are fully entitled to ask about what happens with high velocities and to expect intelligible answers. A more penetrating look at what we are doing suggests that in the manner which has prevailed since Newton, certain phenomena have been put on a special footing: motion relative to the ether is in fact absolute motion and the Lorentz contraction, if real, is an absolute contraction for which observers everywhere would find the same magnitude. It may be that the strange result which has appeared at high velocity arises from the use of this human concept of absolute motion but if we are to abandon it, some firm new basis will be needed on which to build a system of dynamics and, because the phenomena of light are inextricably involved in it, the new system must also be one of electrodynamics.

In the development of the new dynamics, Albert Einstein, as everyone knows, was the leading figure and the epoch was defined by the publication in 1905 of his special or restricted theory of relativity. It is rather easy to assume that Einstein's work was a direct consequence of the Michelson-Morley experiment but in fact that does not seem to be the case. (4) The famous paper of 1905 stems from the classical physics of Clerk Maxwell and Hertz. There is a reference to unsuccessful attempts to measure motion relative to the ether but no detail appears to show whether or not Einstein had the Michelson-Morley experiment specifically in mind. It was not, after all, the only ether wind experiment to have been tried, it was the best.

As so often happens when great changes in scientific outlook are taking place, a group of leading thinkers was involved. One has heard some fine talk about the advancement of science, step by step with inevitable logic ground out by dispassionate impartial scientific observers. Of course scientists work as logically and impartially as they can but a big step can only be made with penetrating vision and a refusal to be hampered by what *seems* to be logical. Lorentz, or for that matter Poincaré, might have anticipated Einstein's relativity had it not been for intellectual barriers which they could not cross.

Though the Michelson-Morley experiment caused scientific turmoil, the upheaval was not quite violent enough to drive Lorentz over the barrier into a world with no ultimate standard of rest. From reading their publi-

---

4. This kind of assumption often arises from uncritical acceptance of what someone, apparently authoritative, reported the original investigator as doing or saying. It pays handsomely to read original texts. In this particular case I have to thank Prof. Bondi for pointing out that Einstein was probably not strongly influenced by the Michelson-Morley experiment.

cations it seems fair to say that Lorentz was a brilliant formal thinker whereas Einstein had a powerful visual imagination. For Poincaré it was unthinkable that geometry could be treated as an experimental science. How remarkable that one of the most powerful intuitionists should have been so bound by 19th century positivism. In his search for a better epistemological basis for dynamics Einstein recognised the barrier and elected to cross it; there his achievement began.

At first sight this new world will seem rather strange and, following Einstein, we shall need some kind of mental compass to guide us. The obvious feature is the constancy of the velocity of light. There is experimental evidence for it and a theoretical basis in the classical theory of Maxwell. Here is something pointing away into the unexplored new world and comfortingly back to the old. Let us see what can be learned about observers working with light. From now on observers are going to be important people, indeed for cosmologists, the only important people, so it will be as well to start with a somewhat restricted class of observers making simple observations.

The velocity of light has the same value for all inertial observers. Inertial observers it will be recalled, are those moving with uniform relative velocities. Observers on the earth can fairly be considered as inertial because the speed of a laboratory on the earth does not change significantly in magnitude or direction during the short time of transit of the light in the Michelson-Morley apparatus. [5]

Newton's principle of relativity stated that all inertial observers are dynamically equivalent. There is no dynamical experiment conducted wholly within his frame of reference which will reveal an inertial observer's velocity. [6] In other words, all inertial observers get the same results from the same experiments or, in yet other words, the laws of dynamics are the same for them all.

---

5. I suppose it would be proper to say that the speed does not change significantly relative to some Galilean coordinate system. For a discussion on what Einstein and others have meant by inertial frames of reference see DINGLE, H., On Inertial Reference Frames, Science Progress, Oct., 1962, P. 568 ff.

6. Take the case of a passenger in a well sound proofed airliner in a steady, straight and level flight. A typical dynamical experiment like dropping an object from roof to floor of the cabin will not reveal the motion of the aircraft, it drops vertically just as if the aircraft were stationary. However an acceleration due to change of course or speed, which makes the observer non-inertial, would be revealed by such an experiment. If looking out of the window, all he can see is another airliner also in steady, straight, level flight, but with a different velocity, he can tell by observation that there is relative motion between them, and, given suitable instruments, its magnitude, but in this way neither observer will get a figure for his own velocity through the air.

Einstein's restricted or special principle of relativity, so called because it is confined to inertial observers, goes further in that it takes into account also the motion of light. It may be formulated thus: all inertial observers are equivalent. ([7]) This is by no means the only formulation and as it is sound practice to take notice of what the originator of a proposition said about it, two quotations in translation from Einstein follow: —

> "If a system of co-ordinates K is chosen so that, in relation to it, physical laws hold good in their simplest form, the same laws also hold good in relation to any other system of co-ordinates K' moving in uniform translation relatively to K." ([8])

> "If, relative to K, K' is a uniformly moving co-ordinate system devoid of rotation, then natural phenomena run their course with respect to K' according to exactly the same general laws as with respect to K." ([9])

If all inertial observers are to find the same velocity for light a new law for compounding velocities is needed. As the Michelson-Morley experiment shows, the Newtonian rule for velocities in the same line, $v_{ab} = v_a + v_b$, leads to an absurdity when one of them is the velocity of light. The new rule must be equivalent to the old for v_a and v_b, small compared with c, and give a result consistent with observation for values which are substantial fractions of c. Not unexpectedly, the Lorentz transformation enters into the derivation of the new rule which has the form: —

$$v_{ab} = \frac{v_a + v_b}{1 + \dfrac{v_a v_b}{c^2}}$$

For small values of v_a and v_b this rule reduces to the Newtonian rule in effect, since $v_a v_b / c^2$ approaches zero. For large velocities this factor approaches unity; thus when v_a and v_b each have the value $\frac{1}{2}c$, v_{ab} is $0.8c$ for velocities either of approach or of recession. By doing the arithmethic for values like $0.9c$ it is to be seen that v_{ab} comes very close to c but does not reach it unless one or both of v_a and v_b is c.

There is a point to bear in mind here; all this has to do with the velocity of light in vacuo, the wave velocity with which the group velocity coincides. This does not mean that a group of waves, a flash of light, would

7. As Professor Bondi once said, using this formulation in a lecture; it is very much to the point as in the final analysis, all practical observations require the use of light.

8. EINSTEIN, A., *The Foundations of the General Theory of Relativity*, reprinted in Dover S 81 Loc. Cit. P. 111.

9. EINSTEIN, A., Bonanza reprint, Loc. Cit. P. 13.

not reach the observer sooner when he is in motion towards the source of light, nor later when he is receding from it. We are compounding velocities, not travelling times.

It follows from this rule that there will be a limiting relative velocity c for all motions of a physical nature, only purely geometrical velocities will be exempt. Such a conclusion seems just about as unlikely at first sight as the nature of the ether which caused so much trouble before but if we accept it, we shall be able to arrive at a much simpler view of natural phenomena, consistent with observation and not involving a multiplicity of additional hypotheses.

Before following further along the path taken by Einstein, what of the ether? Relativity does not deny the existence of the ether; it is simply not interested. Relativity is concerned with the laws of motion which give a consistent account of the observed behaviour of bodies and of light. If anyone wishes to retain an ether wave theory of light or a corpuscular theory or some combination of the two, he is quite at liberty to do so, provided that it does not predict phenomena which turn out to be inconsistent with observation. [10]

Escape from all these problems, which in the long run can be seen to arise from the hypotheses of absolute space and time, involves an excursion into epistemology; we must find out how we really come to know about natural phenomena, more particularly about the motions of bodies. Upon reflection it appears that we gain knowledge by means of light and since light travels at a finite speed which cannot be surpassed by anything else, we can only know about distant events some time after they have happened. For astronomers this is a commonplace notion; obviously how long ago some phenomenon, observed now, happened depends on the distance. There is no way to observe what is happening at a distance now. In these circumstances it will be as well to examine more closely what we mean by simultaneity. It is by no means certain that we can give an intelligible meaning to *there-now*.

No problem arises about *here-now;* two events occurring at the same place and at the same time are simultaneous in every sense of the word. But the notion of absolute time has accustomed us to thinking that we can unambiguously describe two events widely separated in space as simultaneous. Conceptually the events could be timed by clocks, one at the

---

10. Thus we find Eddington, a convinced and very advanced relativitist still talking about ether waves in 1927 in his book *Stars and Atoms,* A relevant section is reprinted in SHAPLEY's *Source Book in Astronomy, 1900-1950* Harvard, 1960. P. 220.

site of each event and both having their hands in exactly the same position at the same absolute time; then events occurring at the same indicated time would be simultaneous whether the sites of the events were in motion or not. If we are going to do without an absolute time, there is no guarantee that *then* in two separate places has precisely the same significance as with absolute time. The idea of simultaneity in the old sense can be used when the events and observers are all at rest relative to one another but it is much more important to find out what it may mean when, as in the real world, motion is involved. Upon looking into the situation we shall find that events which one observer sees as simultaneous are not necessarily so for others. From this investigation many other interesting consequences will follow, as Professor Bondi has so clearly and refreshingly shown; indeed we cannot do better than to follow him. (11)

Let us consider the case where we have two observers, A and B, in uniform relative motion in which their distance apart is increasing. A can send out a light flash which is returned or reflected by B with no loss of time but, at the moment he does this, he is farther away than when A sent out the flash. The situation can be represented diagrammatically as in Fig. 6.

The motion being purely relative, A and B are completely equivalent and it is only for convenience that in our diagram we represent A as at rest and B in motion. On the paper we use a vertical axis for time and a horizontal one for distance, so that A at rest is represented by a vertical line — he has duration but no motion. B is represented by a line sloping to the right, the direction of increasing distance; obviously the angle between B's line and the horizontal axis depends on his velocity. Both parties have clocks reading their respective proper times (12) so that each can identify events for his own purposes. It will simplify things if at the moment A and B are together at 0, they synchronise their clocks. This is easily done then, for if A sends out a flash of light B sees it at that moment. Now as they move apart let A send out flashes of light at equal intervals t. These can be marked on his line starting from 0. Because of the high speed of light the path of the flash will be at a rather small angle

---

11. The treatment of special relativity which follows is adapted from that developed by Bondi some years ago. His latest version of the subject is to be found in *Relativity and Common Sense*, Heinemann No. 31, 1965. This book contains much illuminating background material and should be read by all students.
12. In cosmology and in relativity the word proper has the sense of pertaining to, associated with the noun property, not the sense of suitable or correc.t

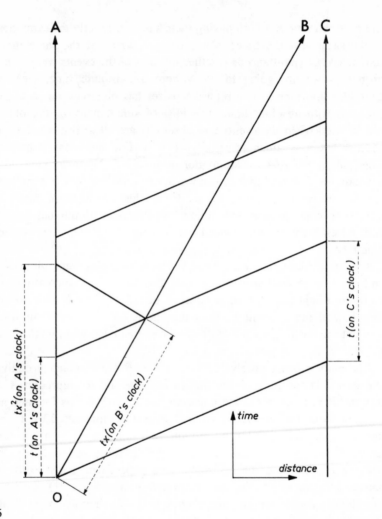

Fig. 6

to the distance axis; light goes a long way in a short time. If the distance between A and B had remained constant then B would receive A's flashes with the same interval t between them as they were sent out, but the whole series of flashes would be received later in time. As however, the distance between A and B is increasing at a constant rate, each successive flash has a greater distance to cover and so the series will be received by B at intervals according to his clock which are greater than the intervals t according to A's clock. Let us represent these greater intervals by tx, where x has a value greater than 1. It is clear that $x = 1$ would be the value for constant distance. We could equally well consider the case where the distance between A and B is decreasing: then B would receive the flashes at intervals which his clock would show to be less than A's intervals t. The value of x would then be less than 1. If we wish x, as an unchanging factor in the physical situation, to retain the same numerical value as for the case of increasing distance, we can represent B's smaller intervals as t/x. When B returns the flash sent out by A at t and received by B at tx, it is received back by A at time tx^2. The time for the flash to go both ways is thus $(tx^2 - t)$ and the time for one way is

$$\frac{t(x^2 - 1)}{2}$$

From this, B's distance at moment t can be calculated in terms of the velocity of light.

The moment when B received the flash, which he read from his clock as tx, is on A's clock, half way between tx^2 and t, that is —

$$\frac{t(x^2 + 1)}{2}$$

A can now find B's relative velocity since in time

$$\frac{t(x^2 + 1)}{2}$$

he has travelled a distance

$$\frac{t(x^2 - 1).}{2} \times c$$

In terms of the velocity of light, B's velocity v then works out as —

$$v = \frac{x^2 - 1}{x^2 + 1} \cdot c$$

For the case of relative motion where the distance is decreasing, our synchronising point 0 would lie in the future. However, we can get around that by putting in another observer C to the right of the diagram, whose line is also vertical for he is at rest relative to A whom B is approaching. C will receive A's flashes at intervals t as they are sent out because his distance from A is unchanging. Obviously the whole series of flashes will be received by him later in proportion to his distance from A.

While finding B's distance and velocity relative to A, we have come across something else of interest. Those two clocks which we so carefully synchronised and which we should expect to remain synchronised, do not. According to B, he received the flash at time tx by his clock, but according to A's clock it was at time $t(x^2 + 1)/2$, and these are not the same. Moreover, this difference is reciprocal, for we can just as easily work out the problem on the basis that B is at rest and A is moving. The same results are found for converging courses such as those of B and C.

When we have two observers in relative motion each sees the other's clock as running slow. In the case of A and C, C's clock readings start off a little later, but the rate of his clock is the same as A's, and they both see B's clock as running slow and he likewise has the same opinion of their clocks. This all seems highly paradoxical but it follows inevitably from the laws we adopted. This effect is found in the physical worlds as well as on paper and it is called the relativity of time.

Having come to the conclusion that our proper time when applied to an observer in motion relative to us is not the same as the proper time at that observer's position, we can ask whether this really matters much, since we know how to calculate the difference and make allowance for it. So long as there is only one meeting of the observers concerned either in the past or in the future, it does not really matter much. A more important question is; what happens if they can meet twice?

A and B met once in the past when they synchronised clocks. For them to meet again at least one of them must be accelerated by a change of course. Let us suppose that B is accelerated. After some time from the original meeting he changes his direction so as to come into coincidence with A again later on. Here we must be careful in our ideas. While B is accelerating he is not in an inertial system of reference and A and B are not equivalent. Accordingly we should not expect the time dilatation, as the slow running of the other observer's clock is called, to be a recipro-cal effect. It makes no difference in principle whether the acceleration is abrupt or whether the change of course is smooth and gradual. In practice as we know from personal experiences, abrupt accelerations involve large forces and are likely to have destructive consequences.

Once he has settled down on his new course, B can again be treated as an inertial observer since he is then in uniform motion relative to A. In figure 7, we represent the situation with A and B as before but we have brought in C going in the opposite direction at the same speed as B, so as to meet A. After passing A at 0, B meets C at P, who later meets A at Q. The meetings at 0 and Q are in the same *place* but Q is later in time. The meeting at P is in a different place and between the others in time. The reason for bringing in C is to avoid the acceleration of B through a reversal of his course in zero time, which would otherwise be necessary to keep our calculations simple.

As before, A sends out a light flash after the interval t and this reaches B at P at time tx by his clock. Purely for convenience we arrange matters so that C arrives at P simultaneously with the flash and his clock can

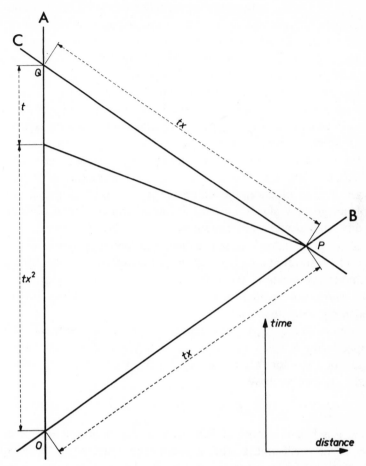

Fig. 7

then also be made to agree with B's. From P, B and C jointly return the flash to A who gets it at time tx^2 by his clock.

As C is moving towards A with the same speed as B is receding, he will meet with A after a time tx by his clock, but this will be only t by A's clock after the flash from P. (Since C's velocity is one of approach, we have to divide by x.) The whole time from 0 to Q by A's clock is therefore $t(x^2 + 1)$, while the whole time for B and C from 0 to Q via P is $2\,tx$ by their clocks and this result is the same as if it were only B who had been suddenly accelerated at P.

It is easily seen that $t(x^2 + 1)$ is greater than $2\,tx$ except in the case where $x = 1$ (which is the case for no relative velocity and so does not concern us). We are thus faced with a disagreement between A and B (represented by C) as to the time interval between two events (the meetings at 0 and Q) in which both parties are perfectly correct in their clock readings. The only difference is that one has stayed put and the other has gone off somewhere else during that interval. This again seems totally against common sense and experience. We may suppose A and B are two inhabitants of some city and they have very accurate and reliable watches which are regulated to agree precisely. B gets into an aeroplane and flies round the world. Our conclusion ought to mean that when he gets back it is later by A's watch than by his. Indeed this is theoretically the case but, in practice, the difference is too small to detect because aeroplanes go very slowly indeed compared with the speed of light; the time difference would be in millionths of millionths of a second.

This "clock paradox," as it is called, only becomes important when it is possible to travel at speeds which are a substantial fraction of the speed of light. Hence we have the story, so far fiction, about identical twins, one of whom is sent off in a space ship and returns to find his brother considerably older than himself.

As we would expect, the relationship between earth time and time on the space ship is given by the Lorentz formula. If we represent earth seconds, or years for that matter, by t and the corresponding space ship units by t^1, then we have

$$t^1 = t \sqrt{(1 - \frac{v^2}{c^2})}$$

Thus if v is half the speed of light we find that a spaceship time unit is 0.866 of the earth unit, so that a year's voyage according to the spaceship's clock would be nearly 14 months by the clock on earth. If v were 0.9 of c then a spaceship year would be about 23 earth years. The difference will obviously grow very rapidly as v gets closer still to c.

The relativity of time is something that transcends ordinary experience and will continue to do so unless one of these days technological advances make the spaceship story come true. None the less there is evidence that time dilatation does happen. Among the many mysterious particles with which nuclear physicists make experiments are some exceedingly short lived ones, having existences lasting much less than a milionth of a second between their production and disruption. It is an extraordinary feat to detect such a particle, let alone determine their evanescent lifetime, but these things are possible and the experimental results are reliable enough. Now it is found that if these particles are formed with velocities ap-

proaching c, their observed lifetimes are much longer, in accordance with
the time transformation shown above. There is also strong evidence of
the same effect with some of the ephemeral particles produced by the
impact of high energy cosmic rays on atoms in the earth's atmosphere.

We have come a long way from the objective simultaneity of Newtonian
mechanics and though the new notions may be rather strange compared
with the very definite idea we started with, the final result is much more
significant and does not introduce any embarrassing conflicts with observations of high speed phenomena. It is now time to consider other effects
of raising bodies to very high speeds, bearing in mind that there is a
limiting velocity c. Clearly the second law of motion, $F = ma$, will no
longer serve in this simple form unless a new interpretation can be found.
A constant force cannot produce uniform acceleration indefinitely, yet
constant F certainly means that the product ma is constant. There is very
good evidence derived from chemistry, that matter cannot be created or
destroyed. This law of the conservation of mass is certainly supported by
the most refined tests that chemists can contrive but after all, chemistry
deals with matter in a rather static way; even violent explosions do not
involve relativistic velocities so perhaps the conservation law does not
apply to matter moving very fast. We should then get a new rule in the
form: —

F = (mass x rate of change of velocity) + (velocity x rate of change of
mass).

A constant value of ma clearly implies this, for since a must decrease
then m must increase. If a body could be raised to the velocity of light
by the application of some force the acceleration would finally have to
become zero and the mass infinite. If the velocity is known the mass of
a moving body can always be found from its momentum, which quantity
can be determined from the impact of the body, like a projectile upon a
target. Even before the publication of Einstein's special theory of relativity
in 1905, it was known that electrons travelling at high speeds possess
momenta in excess of the values to be calculated from their velocities and
their mass as determined at low speeds. [13] This effect in relation to
particles like electrons, protons and even whole nuclei, is nowadays re-

---

13. This had been observed in experiments with vacuum tubes. The velocity of the
electrons could be calculated from the voltage gradient and their deflection by
a known magnetic field gave the effective mass. The phenomenon was well
known to Lorentz and to more experimentally minded physicists like Kaufmann
and Righi, the latter of whom published his *Modern Theory of Physical
Phenomena* in 1904.

garded by physicists as a truism and can be demonstrated by comparatively simple experiments.

The increase in the effective mass of a body with velocity, as revealed by its momentum, can be calculated with the aid of the Lorentz transformation. At some velocity v relative to the observer and the target, its mass m is related to its rest mass thus: —

$$m = \frac{m_0}{\sqrt{(1 - \frac{v^2}{c^2})}}$$

From the effects of its impact on a target its increased mass seems real enough but in what does it consist? Now there is only one thing which is contributed to a body when it is accelerated and that is energy; the kinetic energy of a body increases with velocity, this being energy which is expended in accelerating it. At any given velocity the kinetic energy E of a body is $\frac{1}{2}mv^2$. In the classical view of dynamics energy and mass are quite different but mass can obviously possess energy; perhaps then the extra mass of a moving body is simply its energy which has in some way materialised for the time being. Experiments made with fast moving particles having horizontal trajectories show that the increased mass responds to gravity just like ordinary mass. [14] It seems by every test we can make to be a genuine mass.

If this is indeed the case it may well be that in ordinary matter the mass consists of energy concentrated and locked up in some more or less permanent form. At low velocities the kinetic energy of a moving mass is still effectively $\frac{1}{2}mv^2$ but when we have to take into account the increase of mass with velocity, the energy required to change m_0 to m is measured by the change of mass multiplied by the square of the velocity of light, it looks as if the rest mass m_0 is also energy, as we have suspected. It appears then that we can fairly describe the total energy of a moving mass as the sum of its kinetic and rest-mass energies:

$$E = (m - m_0) c^2 + m_0 c^2$$

which simplifies to $\quad E = mc^2$.

This last equation is the famous Einstein equation for the mass equivalence of energy. [15] Its validity has been demonstrated by many ex-

---

14. In Newtonian theory the inertial and gravitational masses of a body are distinct concepts, having different origins but their magnitudes happen to be the same. See WHITROW, G. J., *The Structure and Evolution of the Universe*, Hutchinson, 1961, P. 93.
15. EINSTEIN, A., *Relativity*, Loc. Cit. Chap. XV.

The Small Magellanic Cloud, occupying 2½° of the sky and distant a little over 200,000 light years. It is connected to the larger cloud by tenuous clouds of hydrogen. The circular image close to it is that of the globular cluster 47 Toucanae which is part of our galaxy and is 20,000 light years from the sun. Mount Stromlo photograph.

M31 The Andromeda Nebula, the only external galaxy visible to the naked eye and, apart from the Magellanic Clouds, our nearest neighbour in space at about 2 million light years distance. Probably our galaxy would present a similar appearance to Andromedan observers as theirs does to us. This is a Lick Observatory photograph.

periments and very spectacularly by the explosions of nuclear weapons. What we regard as a small amount of mass or matter can be converted into what is, by ordinary standards, an enormous amount of free energy.

Since matter and energy now appear to one and the same, either free or tied up in small bundles as the ultimate particles of matter, two further questions arise, of which the first is: does energy, say in the form of light, have gravitational mass? The answer is unequivocally; yes. At once the human tendency to visualise comes into play, trying to produce a picture of an indefinitely long wave train with mass distributed along it, like a length of waving string. This visual model can be developed into a more precise one with mathematical form and it works quite well until it has to account for certain interactions between light and matter, which leads to the second question.

Is free energy infinitely divisible or does it too occur in lumps, atoms of energy as it were? Once again the answer is yes.This explains mysterious features of some interactions of light and matter. In photo-electric phenomena, light incident on certain substances can dislodge electrons from them, producing in appropriate apparatus, an electric current. Photoelectric devices are very widely used, in television, cinema projectors and in industrial control equipment to mention only a few. Now it requires a very small but quite specific amount of energy to dislodge an electron but this cannot be done by allowing the energy of light of any frequency to accumulate. Thus there are substances in which blue light will readily liberate electrons but no amount of red light will do so, even though the total energy supplied may be very much greater.

The explanation is that the energy of electro-magnetic radiation comes in discrete amounts which are called "quanta." [16] (Light quanta have been given, by customary usage, the more special name photons.) To make an analogy, one can set up a steel target of some thickness such that the bullet from a low powered rifle will not penetrate it, and of course, it does not matter how many such bullets are fired at this target, none will penetrate it. However any bullet fired from a suitably high powered rifle will penetrate the target. We can thus think of light as bursts of waves containing just so much energy, according to its frequency. The quantitative relationship between the energy of a photon and its frequency has been found;

---

16. No more than what appears in this need be known about quantum theory for the purposes of elementary cosmology but an informative and entertaining account of it for interested readers can be found in HOFMANN, B., *The Strange Story of the Quantum*, Pelican, A587, 1963.

$$E = h\nu$$

where ν according to optical practice, is the frequency and h is a quantity called Planck's constant. h has the extremely small value of 6.624×10^{-27} erg/sec. so that although ν is rather large, the energy of a photon is very small. For this reason the number of photons involved in the faintest glimmer of light detectable by the eye is so enormous that we get no impression whatever of discontinuity in light.

The mass of a photon can be calculated from its energy and the Einstein equation; for visible light this comes out around 10^{-33}g. For many purposes it is convenient to think of the limited bundle of waves in a photon as a particle. This is very suitable for interactions with matter like photo-electric phenomena, while the wave notion has greater appeal when we look at polarisation and interference.

Towards the low frequency end of the range of electromagnetic radiation where we find radio waves, the energy content of a quantum is extremely small indeed and there are virtually no phenomena that cannot be satisfactorily dealt with on the wave basis, indeed, on the classical continuous wave theory. [17] At the other extreme in the gamma ray part of the spectrum the quantum approach is the more effective so there radiation is treated as essentially particulate. As we have seen, light is more in the middle range where both approaches have their advantages. The modern scientific view is not that light is waves or is particles but that it presents to us wave and particle aspects. While it may be philosophically disturbing to some people that we no longer hope to say what light really is, the price seems to be a modest one for what we gain in the possibilities of exploring the behaviour of nature.

The same can be said for special relativity; it has been well worth while to sacrifice some hypothetical absolutes. To round off the discussion, what about that absolute Lorentz contraction of a moving body with which it started? It is still very much in evidence in a relative way. An observer sees a body moving relative to him as shortened in the line of motion and it is just as real for him as all the other effects calculated with the Lorentz transformation. But an observer on the body we have just described as moving is entitled to see himself at rest and the first observer in motion relative to him, complete with slow-running clock, enhanced mass and suitably foreshortened length. For inertial observers their mutual equivalence is complete.

---

17. DICKSON, F. P., *Of Coherence*, Scientific and Industrial Equipment Bulletin, Philips Electrical Pty. Ltd., Sydney, Vol. 4, No. 9, 1965, P. 15 ff.

From what has so far appeared in the discussion of special relativity, there seems to be a consistent basis for setting up laws of motion without appeal to absolute space and time but it is obviously necessary to go further in several ways. There is something absolute about acceleration in the theory so far and, as we know from observation of the world around us, accelerated motion is much more often found than inertial motion, so we shall have to lead on to some elementary notions of general relativity to deal with that. Meanwhile the relativity of time and length, hence of distance, leaves a disturbing sensation of fluidity about space and time. Clearly what is lacking in special relativity so far and what will be necessary later for general relativity, is some factor in observations which will be the same for all observers without or despite the use of transformations, something invariant [18] to provide a firm tie to the world.

The only factors available are space and time and since neither will serve alone the answer must lie in some combination of the two. Events happen at places at times; nothing can exist in space unless it endures for some time nor can it endure except at some place. The way in which space and time can suitably be combined was brilliantly set out by Minkowski in the idea of *space-time*. [19] Let there be no misconception here; *space-time* is not space + time, it is an indissoluble combination of the two into a four dimensional entity, technically called a manifold or continuum. It is quite impossible to visualise a system with four dimensions all mutually at right angles but it is not difficult to think about it in a consistent logical way and to use it as a theoretical model of the real world. The four dimensions have to be derived from three spatial dimensions and one temporal one in such a way that they are all strictly equivalent. The velocity of light comes into this as the connecting factor because, as we have already seen, it enables us to express time as a distance and distance in terms of time. Consider two successive events somewhere here on the earth, one second of time apart; we say that they happened at the same place and so they did for us; but for an observer at rest relative to the sun, they were separated by a spatial distance of roughly 18 miles, the distance travelled by the earth in that time. In space-time it is legitimate to express extension in all four dimensions in space units or in time units according to which is the more convenient for the case in hand. If that hypothetical observer was on the sun, the

18. CASSIRER, E., *Substance and Function*, Dover Reprint T 50, 1953, P. 249 f.
19. MINKOWSKI, H., *Space and Time* reprinted in Dover, S 81 Loc. Cit. P. 75 ff.

two events would be distant from him by 93,000,000 miles or 8.3 light minutes.

The unvisualisable four dimensions at right angles can be very simply handled by an extension of Pythagoras' theorem. The method of specifying the position of a point on a plane is very well known, using two axes at right angles, conventionally called x and y, with 0 the point of intersection, called the origin of the co-ordinates. Then the distance of any point in the plane from 0 is given by:

$$s^2 = x^2 + y^2.$$

In three dimensions the procedure is equally simple. The position of any point in a room is specified by measurements starting from a corner where the junction of two walls and the floor serve as coordinates. The distance of a point from the corner is found from the relation —

$$s^2 = x^2 + y^2 + z^2$$

Now a fourth dimension which may be called t can be brought in:

$$s^2 = x^2 + y^2 + z^2 + t^2$$

Mathematically this is perfectly simple and consistent with the rules for manipulating the symbols. There is no magic about four dimensions, or fourteen for that matter; if some phenomenon requires more than three dimensions to give expression to its character then we simply use them. The only remarkable feature of three dimensions of space is that it is the system of our own experience in the everyday world and there is no logical justification for insisting that it is the only possible system. A couple of generations ago there was quite a fashion for mystical speculation about the "Fourth Dimension" which was pressed into service to explain spiritual experiences and miracles but we should take no notice of that kind of mental exercise, for all that is being done here is to find convenient ways of representing things. Scientific models are for use until something better can be devised and are in no circumstances articles of faith.

Since t is not just another spatial dimension but has to represent the time aspect of space-time, it must be put into a form mathematically compatible with the other units. Although it is to be combined with and treated in the same way as space, we do not say that time is identical with space. In space-time the compatible measure of time is symbolised by ict. There is nothing remarkable about ct, that is clearly a distance, but it may seem strange that i which is $\sqrt{-1}$, should appear. There are several reasons for bringing in i, including mathematical convenience and it must be remembered that multiplication by i does not alter the magnitude of a quantity, for i is not a number; it is an operator and signifies a change of direction by one right angle when the quantity is a distance along a co-ordinate. Multiplication by i^2 brings about a shift of two right angles, thus a reversal of sense as indicated by —1. The advantage of this notation for time co-ordinate will appear as the discussion proceeds.

A point in ordinary Euclidean geometry has position but no magnitude and as this geometry is timeless, nothing is said about its duration. Similarly an instant has position in time but no duration. In space-time the equivalent of both is a *point-instant* to which Minkowski gave the name *world-point*. As determined from the origin, a world-point has position on all four axes but no magnitude on any of them. An Euclidean point at rest in space but having duration in time would appear in space-time as a straight line parallel to the time axis. In the same way a body having duration has to have a world-line in space-time.

For the study of motion it is not much help to think about the world-line of just one body in all space-time. Of course any number of world-points and lines can be located relative to the origin of coordinates, but there should be some convenient way to express the relations of neighbouring points without going back to the origin. Where dx stands for a very small portion of x, the distance along the x axis of a point adjacent to a point of distance x can be expressed as $(x + \mathrm{d}x)$. Elaborating this for four axes, we come to the rather clumsy set of symbols;

$$(s + \mathrm{d}s)^2 = (x + \mathrm{d}x)^2 + (y + \mathrm{d}y)^2 + (z + \mathrm{d}z)^2 + (t + \mathrm{d}t)^2$$

Fortunately this can be taken from the original s^2 leaving us with the *Sloppy* desired part, ds^2 and, putting in ict as the proper replacement for t, we have:
Can get this ←*directly*

$$\mathrm{d}s^2 = \mathrm{d}x^2 + \mathrm{d}y^2 + \mathrm{d}z^2 - c^2\mathrm{d}t^2.$$

As it does not matter which end of ds we start from the equation can be put in the form:

$$\mathrm{d}s^2 = c^2\mathrm{d}t^2 - (\mathrm{d}x^2 + \mathrm{d}y^2 + \mathrm{d}z^2).$$

The quantity ds, corresponding to the distance between two neighbouring points in space, is called *interval* and it is the invariant quantity for which all inertial observers find the same value; the invariant common feature of observations that we have been looking for in the world of special relativity.

Interval is naturally not the same as distance, for we can have interval between two events happening at the same place; then dx, dy and dz are all zero and the interval reduces to the temporal separation. In this sense a point can have interval between itself, as it was and as it was later. For two reasons interval is always positive; when interval reduces to time alone it is always getting later; when motion is involved interval could be negative if motion along the x axis, for instance, made dx greater than *c*d*t*, but that cannot happen because it would mean a velocity greater than that of light. The mathematical manoeuvre of putting *i* into the time factor gave expression to these aspects of interval.

In another and seemingly paradoxical way interval differs from distance. Everyone knows from Euclid that the shortest distance between two points is a straight line, but even though the space-time of special relativity may be as flat as Euclid's space, this statement is not true for space-time; in fact an interval which is a straight line is the longest possible line between two world-points. This kind of idea is one which is instinctively rejected and perhaps even resented because it is so unfamiliar and contradictory to experience; it is the old problem of trying to visualise what can only be thought about.

It is easy enough to see how this comes about mathematically. In pure space ds^2 is the sum of all the space elements, dx^2, dy^2 and dz^2. In space-time ds^2 is the difference of the time element c^2dt^2 and the sum of the space elements. There is a natural tendency to enquire whether this strange character of space-time is a matter of mathematical trickery or whether it is a fair representation of the facts of observation: it is quite certainly a fair representation. The difference of squares basis also leads to the conclusion that along the path of a ray of light $ds^2 = 0$, in other words there is no interval. In the case of a light ray travelling along the x axis, dy^2 and dz^2 are 0 and we have:

$$ds^2 = c^2dt^2 - dx^2$$

but, for light, $dx^2 = c^2dt^2$ since $dx = cdt$

Therefore $\qquad ds^2 = c^2\,dt^2 - dx^2 = 0$

At this stage it will be instructive to consider a space-time in which one or two of the spatial dimensions have been dropped for ease of handling. Let us think about a space-time with only two dimensions of space, a flat world in which there are points, lines and areas but no volumes, all spatial extension is confined to a plane and the time axis is at right angles to the plane. Naïvely, we can see a horizontal plane with the time axis sticking up vertically out of it to the future and downwards to the past, so that *now* is where the time axis intersects the plane. Something can exist only where it is and in this model we have confined *where* to the

Spatial plane (Fig. 8).

At the origin 0 where the spatial and temporal coordinates meet is *here-now* and there are two regions of *there-then,* the future above the plane and the past below it. But for us observing from *here-now*, not the whole of those two regions is accessible. Bearing in mind the role of the

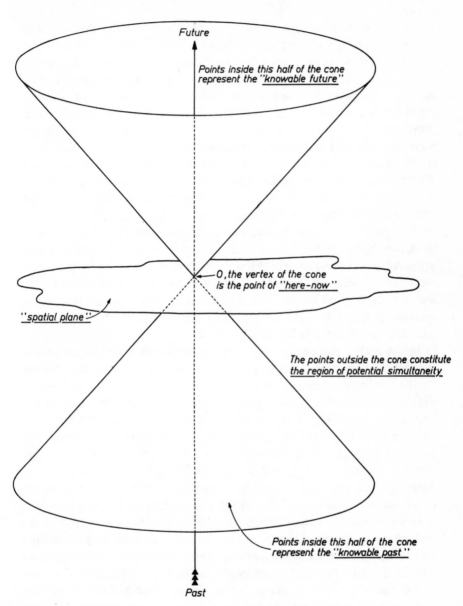

Fig. 8

velocity of light, it can be understood that there may be events about which we never could have or will be able to know because that would involve signals travelling faster than light. The track of a light ray which started its journey in the past and passes us at 0 is represented in this limited space-time by a straight line with a certain inclination between the time axis and the plane, which is determined by its velocity. The projection of this line upon the plane indicates the spatial direction of travel. If light rays are made to pass through 0 from all directions, their set of tracks will describe two cones with their vertices meeting at 0. [20] Within those light cones are the regions of space-time where it was or will be possible for an observer at 0 to know about events. Within the light cones a body at rest has a straight world-line parallel to the time axis, while a body in uniform motion has a straight world-line at some angle to it, less than the angle of the rays of light. A body which is under acceleration has a curved world-line with a shape depending on the particular law of acceleration under which it is moving. In accordance with the rule about acceleration never producing a velocity greater than c, no segment of a curved world-line may have a greater inclination to the time axis than the light tracks defining the cones.

In the region outside the cones interval is purely spacelike and, since no temporal significance can be given to things there, it has become known by the rather combersome name of the *region of potential simultaneity*. Much speculative thought has been devoted to questions of the existence and nature of things in this unobservable limbo without reaching any satisfactory conclusions, though it seems that there must be things out there. In principle, each observer has his own pair of light cones, but where a number of observers are grouped together, to all intents and purposes their light cones coincide for the time being. Another group of observers some distance away, say on Pluto, have an 0 which does not coincide with ours in this lecture room, however the two pairs of cones overlap, for in the past there have been events observable by both groups and in the future there will be others. Thus the outburst of a supernova star in the Andromeda Nebula might be such an event. But this also means that there will be events within the Plutonians' light cones that will be outside ours, where the two sets of cones do not overlap. For them these will be quite ordinary events but for us unknowable and even conceptually unlocatable.

---

20. As can be seen from Minkowski, note 29 above P. 77 f and P. 83 ff, the formal derivation of light cones is rather more complex than the oversimplified version given in the text.

No doubt you will have realised that *here-now* at 0 is a point-instant and for an observer whose existence continues *here-now* shifts along the time axis, which of course is implied by the fact it is always getting later and so his light cones ride steadily up the time axis. Thus things in the region of potential simultaneity become knowable to him as past events where information about them can reach him at the speed of light. Thus an event, say the eclipse of one of Jupiter's moons which may be happening *"now-ish,"* can be observed by him later when his light cones have shifted and light from Jupiter containing information about the eclipse can reach him. Here is further confirmation if we still need it for the $-c^2dt^2$. There is no need, in principle, to move on the spatial axis but motion along the time axis is inevitable and in one direction only. The arrow of time flies only one way. Of course, as a relativist, you may just as well say that 0 remains fixed and time flows by bringing future events nearer and taking past ones further away. If a final choice is to be made it will be on metaphysical grounds.

In this little excursion we have slipped quietly from a space-time with only two spatial dimensions into one with three and without mentioning it; our light-cones have become unvisualisable four dimensional cones, but the concept seems to have been adopted smoothly enough. Now consider the world-lines of two observers *here-now* at 0; one will remain stationary, describing in space-time a straight line parallel to the time axis while the other will take an aeroplane ride round the world. The world-lines of observers at rest or in uniform motion may never intersect, or only once, but if there is acceleration, more than one intersection is possible. The observer who took the aeroplane ride and rejoins his friend has travelled quite a long way in the spatial sense and some way in time also, but the one who stayed at home has travelled in time only. The end points of the interval are the same for both observers because in each case they are together at a point-instant, but the observer who stayed home must have travelled further in time; his interval has the magnitude cdt while the other has interval $c^2dt^2 - (dx^2 + dy^2 + dz^2)$ where t_1 and t_2 are their clock readings of elapsed time. It follows that t_2 must be less than t_1. This is in agreement with the result previously obtained with the use of Lorentz's transformation when it appeared that the traveller's clock ran slower than the stationary one so that on his return he found that less time had elapsed for him than for his friend. Thus along two different lines of thought we have come to the same proposition and if we care to make the calculations the results will be found to agree in detail as well as in principle.

⊛ longer interval accomplished in time t_1

✱ shorter interval accomplished in time t_2

Ah but my Computations people say
Have squared the year to human compass, eh?

CHAPTER 10

SPACE, TIME AND RELATIVITY

So far we have taken it for granted that space-time is flat in the same sense that the space of Euclidean geometry is flat. The terms flat and curved as applied to space and space-time are technical terms with defined meanings and are not intended to present visual images as the same words do when applied to material objects in the ordinary way. Euclidean space is flat because a figure, a triangle for instance, has just the same shape however big it may be; regardless of how long the sides may be, all equilateral triangles have the same angles at the corners and the sum of those angles is always two right angles. There is no limit to the size of a figure; flat space is infinite or "open". It can also be described as space of zero curvature.[1]

In curved space the properties of figures vary with size. We can think about both negative and positive curvature and the negative sort which is so called by convention, makes a convenient starting point. Suppose we construct a very large triangle and find that the sum of the angles is less than two right angles; then we construct a still larger one and find that the sum of the angles is smaller than before. The space in which these triangles are being constructed has negative curvature. It does not matter in the least that you have no experience of triangles with this property, we are dealing with ideas now, not things. Bearing in mind the dangers of visualisation, you can construct triangles on a limited scale that do behave in this odd way. Take a strip of metal about half a metre long and bend it

---

1. From the purely mathematical point of view, the geometry of curved spaces originated in the impossibility of proving the validity of Euclid's postulate of parallels. Mathematicians in the 18th and 19th centuries, notably Bolyai, Gauss, Lobatchewsky and Riemann, succeeded in constructing logically consistent geometries in which there is no postulate of parallels. This proved to be possible only in spaces of hyperbolic (negative) curvature and spherical or elliptic (positive) curvature. For readers interested in this subject an historical and critical account of its development is given by BONOLA, R., *Non-Euclidean Geometry*, Dover reprint S 27, 1955. For a very pleasantly expressed view of spherical space see EDDINGTON, A. S., *The Expanding Universe*, reprinted in Munitz, Loc. Cit. P. 320 ff.

into an arc of a circle with a radius of about two metres. Using it as a ruler draw a small equilateral triangle with sides curving inwards and say 5 cm long. Here the sum of the angles differs negligibly from 180°, but now make one with sides half a metre long and it will be seen that the sum of the angles, as measured with a small protractor at the vertices, is rather less than 180°. That curved ruler enables you to simulate, in the space of everyday experience which is near enough to Euclidean, the construction of triangles in hyperbolic space. The triangles we can measure on the local astronomical scale of a few light-years on the sides are Euclidean so far as we can tell but, on the larger scale of a few thousand million light-years there *might* be a departure indicating that the space of our universe has a slight negative curvature. Whether or not space is "really like that" is a question to be settled by observation. Negatively curved or hyperbolic space, as it is also called, is open and observations showing that our space is so curved would necessarily mean an infinite universe.

In a space of positive curvature, the angles of a big triangle add up to more than 180°. Dealing with triangles like that has been an everyday affair for thousands of observers in the course of their occupations as navigators upon the sea and in the air. The surface of the earth is a curved space of two dimensions. While small triangles on this surface do not differ appreciably from Euclidean ones, large triangles certainly do. Looking at a globe map of the earth it is clear that, in a triangle formed by two meridians of longitude and a quadrant of the equator, each angle is a right angle and their sum is 270°. On a spherical surface the lines marking the shortest distances between points, corresponding to the straight lines of Euclidean geometry are arcs of great circles and are technically called geodesics.

A spherical surface has several interesting properties. It has constant curvature since the shape is the same in every direction and it has a finite area which can readily be calculated if the radius is known but two-dimensional observers, confined to the surface, could not know the radius directly. That involves another dimension quite outside their experience but they could postulate a third dimension and calculate the radius in it from the behaviour of various sized figures. Though finite, the surface of a sphere is unbounded. A traveller who goes far enough in what he takes to be a straight line but which we three dimensional people see to be a curve, will arrive back at his starting point without crossing any boundary. Because the surface is finite there is a limit to the sizes of figures which can be drawn upon it, as can be neatly demonstrated with circles. Taking the earth as a perfect sphere and starting at the North

Pole we can construct circles of progressively increasing radius. When the radius is more than a few miles, it becomes apparent that the ratio of circumference to radius is less than 2π. The greatest circumference is at the Equator, where the ratio is 4 instead of 6.28 . . . . . , and, for still longer radii, the circumference decreases. According to the flat space rules we can say of two concentric circles that the one of greater radius and consequently greater circumference, wholly contains the smaller. On a spherical surface, once past the equator, we have to say that even though the circumference is smaller, the circle with longer radius is the greater and wholly contains the other. That this is really so may be seen by imagining the radius from the North Pole to the Tropic of Capricorn swung round the earth a full 360°, when it will cover in turn every point down to the Tropic. Taking the final step, it is obvious that the longest possible radius reaches to the South Pole where the circumference of the circle has contracted to a point but the circle includes the whole area of the earth.

Another example of a space with positive curvature in two dimensions for which a simple geometry can be devised is an elliptical space, resembling the shape of a Rugby football. This does not have constant curvature in both dimensions but two-dimensional observers could still determine its properties, marking it out, if you please, in circular parallels of latitude and elliptical meridians of longitude.

The idea of positively curved space of three dimensions can now be understood by analogy. While it was convenient to look at two dimensional space from the viewpoint of beings with three dimensional experience, there was no logical necessity to do so. Likewise curved space of three dimensions does not require a fourth spatial dimension into which to curve. There is no fourth dimension in physical space. However many dimensions we may please to imagine for mathematical convenience in handling phenomena, this has nothing to do with physical space. As a matter of fact both non-Euclidean geometry and dimensions in excess of three were used in scientific thought before their application in the theory of relativity. In positively curved space where the angles of large triangles have a sum greater than two right angles, space is finite or closed but unbounded so that a long enough journey in what seems to be a straight line brings us back to the starting point. If the space has constant curvature it is called, by analogy and by analogy only, spherical space. No three dimensional being can find a unique centre in it, any more than a two dimensional being can find a unique centre on the surface of a sphere. Any point serves just as well for a centre as any other and any observer anywhere is just as entitled to say that he is at the centre of things as any other. That statement of Nicholas of Cusa, "The centre is everywhere and the circumference nowhere," is exactly true of spherical space.

In this kind of space sufficiently small spheres differ insignificantly from Euclidean spheres, but on the large scale it is found that the surface of a sphere is less than $4\pi r^2$. There is a sphere of maximum surface area for which r is $2\pi R$, where R is the so called radius of space; when r amounts to $4\pi R$ the surface shrinks to a point. The quantity R is expressed as a length (in the order of thousands of millions of light years in relativistic cosmology) and can be used to calculate the volume of space, but it is a mistake to try to visualise it in this way. A much more satisfactory idea of R can be obtained by thinking of it as a characteristic length or scale factor which contains information about how big things can be. Reverting for a moment to the surface of a sphere as a space of two dimensions, observers strictly confined to that space could evaluate R, which we see as the radius of a three-dimensional sphere, as a purely theoretical quantity giving measure of sizes in their world. Thus it would determine the length of the longest possible equivalent of a straight line which is a great circle, and the size of the biggest possible triangle. In this sense as a scale factor, R can be used in Euclidean space when it tells us that there is no limit to size and that areas and volumes are exactly according to the second and third powers of r, however great r may be. Similarly for hyperbolic space R tells us that the surface of a sphere is greater than $4\pi r^2$ and that when r is very large, comparable with R, the surface increases very rapidly with r. R is often found written as $R(t)$, signifying its value at some particular epoch for, as in the case of an expanding space, R may vary with time.

Once again the question whether the space of the universe can be "like that," meaning positively curved, is one to be settled by observation of very distant objects, when a departure from Euclidean geometry would make it possible to calculate the scale factor R.

Thinking about the curved lines, the geodesics, which are the equivalents in curved space of straight lines in flat space, brings us back to the subject of interval and the coordinates in which it is specified. We tend to think of positions in space as being identified on a grid or mesh, a set of squares for two dimensions and a cubical lattice for three. Then ds may represent the distance of a point from the nearest mesh point or corner. If the space is locally curved in some way, ds will not be a short *straight* line. Similarly, ds^2 will represent a small curved element of surface which gives a better description of the shape of space in that vicinity. It is for this reason that metrical equations are usually left in terms of ds^2 rather than ds. ([2]) Of course positions on a mesh can be identified just as well whether the mesh be square or distorted. There is no need to lay out a fishing net flat and true in order to count up a particular knot: in terms of the mesh we can find x and y, dx and dy, however the net may be draped and ds^2 will contain information about the draping in a particular small region. If the net is suspended in the water from floats and has an overall curvature caused by the current, one might expect to be able to evaluate a quantity analogous to R but it would be a difficult and misleading analogy to pursue. There would be altogether too much temptation to think of R as representing ordinary curvature as seen by reference to a dimension outside the mesh. On the other hand if we take ds^2 as being a small but not infinitely small element it should reflect the overall shape of an unbounded mesh if we observe it at a number of positions. Something like this is implied in the relation between ds and R in the metrical equations used by cosmologists to express the characteristics of space and space-time in their models.

The term metric, intimately connected with the idea of measurement etymologically as well as in significance, conveys that the kind of geometry now being considered involves questions of size as well as of shape. We have to think of R as being the same everywhere in the universe but generally changing in time and consequently determinations of ds at different times will reflect the change in R. Conversely observed changes in ds would lead to a knowledge of how R changes. Perhaps to a purist cosmologists may be guilty of a mathematical solecism here. Properly speaking the d in ds should signify a true infinitesimal but in cosmology we only mean the ds is cosmically small; it might well be large compared with the diameter of our solar system.

When looking at cosmological texts you will often come across metrics, some of which appear rather complicated. While it is not necessary for you to be able to manipulate one of these equations, there is some information which you can glean from them, the general nature of the messages which they are trying to tell you. They are of course, developed forms of the equations for ds^2 which we have already seen and are sometimes expressed in the same rectangular or Cartesian coordinates which we have used. More often, however, they are expressed in polar or spherical coordinates which are mathematically convenient and compat-

---

2. ds is also know as the line element and this term is often, though a little loosely, applied to ds^2 with the implication that the equation should be reduced for ds.

ible with the way distant objects are measured, in terms of a radial distance and two angles. ([3])

As you know, the position of a point on the surface of the earth can be precisely specified in terms of two angular measures, latitude and longitude. As all points on the surface of a sphere are equivalent some convention must be adopted to provide an origin and nowadays we choose the intersection of the meridian of Greenwich and the Equator as the origin; latitude 0, longitude 0. Looking out from the centre of a sphere marked in parallels of latitude and meridians of longitude we would see the surface covered with a roughly square mesh all at the same distance, the radius of the sphere, from the centre. You can now imagine an endless series of concentric spheres in which the radius r increases in suitable shapes. By finding r and two angles the position of any point in three dimensional space can be specified. These are polar coordinates with the exception that geometers have adopted a slightly different convention from geographers in that the angles are measured counter-clockwise from the origin and go the full 360° for each of the angles. The measures are also conveniently expressed in radians rather than degrees.

Our earlier equation for interval can now be put in polar coordinates to give the same information in the form ⊗

$$ds^2 = \left\{ dt^2 - \frac{dr^2 + r^2 d\theta^2 + r^2 \sin^2 \theta \, d\varphi^2}{c^2} \right\} c^2$$

in which the trigonometrical operations make the angular measures look like distances in terms of r.

It is customary in setting out metrics to get rid of the c^2 by choosing a coordinate scale in which c becomes unity, distances being correspondingly expressed in terms of time. R is now brought into the equation as the scale factor determining the spatial element r (or dx, dy and dz). Because we have become accustomed to non static models the form $R(t)$ is used, signifying that R which is the same everywhere in the spatial sense, changes with time. Unless the context indicates otherwise you can take it that R increases in some way with epoch as we are usually considering an expanding model. ([4])

---

3. There is a variety of coordinate systems which have been devised to serve the ends of mathematical and physical investigations. Mathematically oriented students will find a convenient summary of coordinate systems and formulations of standard equations appropriate to them in MOON AND SPENCER, *Field Theory Handbook*, Springer, 1961. Spherical coordinates appear at P. 24.
4. Of course as soon as R becomes a definite function of time we are committed to a specific model. Thus in both de Sitter's and Bondi's R is an exponentially increasing function of time.

⊗ $ds^2 = c^2 dt^2 - \left(dr^2 + r^2 d\theta^2 + r^2 \sin^2 \theta \, d\varphi^2 \right)$

It is useful to think of the space mesh as having a particle at each junction; then for an expanding model you can gain a definite notion of the expansion of the material system including the velocity/distance relation and the growth of the space-mesh.

And then quantity k comes into the equation to express the kind of curvature involved. It has the conventional values $+1$ for positive or spherical curvature, 0 for flat space and -1 for negative or hyperbolic curvature. These particular values are convenient for purposes of manipulation and other values which might be arrived at can always be reduced to these by a suitable choice of unit for R. Noting that k is attached to the part of the equation containing the spatial factors you will conclude correctly that it relates to the curvature of three-dimensional space, not of space-time. The latter involves R as well as k and there is always the disguised c^2 to be taken into account in that case. The cases $k = 0$ and $k = -1$ permit of infinite spaces but for $k = +1$ spaces must be finite.

We can now set out a metric incorporating all these features, such as

$$ds^2 = dt^2 - R^2(t)\frac{(dr^2 + r^2d\theta^2 + r^2\sin^2\theta d\varphi^2)}{(1 + kr^2/4)^2}$$

When for flat space where $k = 0$ the part under the line vanishes. However it does not follow that flat space and flat space-time necessarily go together. Thus in some system the space-time may be curved but a space chosen from three of the four coordinates may be flat.

In general relativity we have to seek an equivalence between accelerated observers, so it will be necessary to think more about what acceleration means. In special relativity it appeared that the Lorentz transformations make it possible to convert observations from one frame of reference to another for inertial motions but that the world-line of an accelerated observer comes out as curved and no conversion is possible. To put it in another way, it is always possible to transform away velocities but not accelerations. Inertial observers have to appeal to causes for accelerated motions, known as forces, acting in the direction of the acceleration produced. For inertial observers these forces are real, just as acceleration is.

It is for this reason that in the classical Newtonian system force is the measure of the acceleration produced in a given mass. This is quite straight forward as long as an acceleration is produced, but questions can arise in other cases. It will help to clarify our ideas if we look more closely at one of these. Consider a mass whirled round at the end of a cord: we will all agree that if the cord be released the mass will fly off tangentially. Many people will tell you that the tautness of the cord is evidence that there is a centrifugal force which holds the mass away from the centre and

NGC7148, a distant spiral photographed with the 74 inch telescope at Mount Stromlo Observatory in red light with an exposure of 2½ hours.

The distant spiral galaxy NGC1566. The sharp white patches are the overexposed images of foreground stars in our own galaxy. This photograph was made with the 74 inch telescope at Mount Stromlo Observatory and required an exposure of two hours in red light.

causes it to fly off when the cord is released. (^{5}) Yet a little thought will show that the mass is continually being accelerated towards the centre of rotation, otherwise it could not continue to move in a circle, that is, always at the same distance from the centre. If we let the cord go, the mass will no longer be so accelerated and, having at every instant a velocity in a direction tangential to the circle, will without any applied force at all, fly off along that straight tangent line.

Thus we are led to postulate a centripetal force always accelerating the mass towards the centre of rotation and therefore acting towards it. There does not seem to be any real need for a centrifugal force and if anyone insists upon having it, then it must be a frustrated force producing no acceleration in the direction from centre to mass. Let us be quite clear about this: centrifugal force is a fiction and nothing more. The only useful purpose it can serve is as a theoretical reaction to centripetal force, maintaining balance.

Now the commonest example we have of a frustrated force is in weight. On this earth weight can be experienced or measured only when gravity cannot produce an acceleration towards the centre of the earth, which is most of the time. A body falling freely towards the earth loses no mass (theoretically it gains a little) but it exhibits no weight. Until recently this was, from the human point of view an academic or at the most, very short lived situation, but now we have astronauts being weightless for days at a time. Perhaps we should, having regard to the possibility of frustrating forces, say that they produce or tend to produce acceleration and that we can measure a force in terms of the acceleration it would produce if it could. This is usually done in terms of a counter force such as the tension in a spring balance. But none the less, as we shall shortly see, gravity does not behave in all respects like other forces, tension or pressure for instance: the spring balance works at equilibrium, but when gravity can produce an acceleration its behaviour is odd. Perhaps there is a way of treating gravity, like centrifugal force, as no force at all.

In the observable universe gravitation is by far the preponderant origin of uniform acceleration but its cause has always been a complete mystery, though its effects are well enough known. In the Newtonian system of dynamics it was necessary to appeal to a force of gravity and, unless there is some medium through which it is to operate, action at a distance must

---

5. Now and then one reads a report of an accident with rotating machinery in which it is stated that the fly-wheel or some other rapidly turning part has burst through centrifugal force. What indeed happened was that the metal of the wheel was not strong enough to constrain its parts to a circular path.

be involved; neither of these alternatives is particularly pleasing. There are several features about the force of gravity which are shared by no other kind of force. Upon reflection it will be seen that these are such as to give it a highly artificial character and it would be an intellectual relief to find a better concept.

The force of gravity does not fit the definition of force given above because, as Galileo showed, it produces the same acceleration in all bodies, regardless of mass; no other force behaves like that. Gravity is an extremely weak force. This is not immediately obvious to us because we are associated with a body of very large mass, the earth; the question is not how some given lump of matter can weigh so much but how it can weigh so little. Gravity is enormously weaker than electric forces which follow the same inverse square law for distance. If we determine the electrical force between a proton and an electron due to their having opposite charges, it will be found to have a magnitude something like 10^{40} times their mutual gravitational attraction. Here another strange feature of gravity appears: the force is always one of attraction, never of repulsion as can happen with electric and magnetic forces nor can a body be gravitationally neutral. There is no possibility of depriving a body of its gravitational field.

As far as our observations go the gravitational field of a body, as mentioned in Chapter 4, is isotropic. All we have learned about gravity is in terms of bodies moving rather slowly and apart from the field of other bodies not affecting its field, we can reasonably expect to discount the effect of a body's motion on its own gravitational field which we assume simply moves with it. But there is a danger in taking this notion very far. Since a gravitational field extends throughout all space, however weak it may become, we are assuming that changes in it move in unison with the body causing the field, even at great distances. In other words, changes in the field can be propagated instantaneously. Even the most modest acquaintance with relativity will raise the question; how does the field at a great distance know that its causative body has begun to move? If the action is instantaneous it means that there is a means of conveying information faster than the speed of light. Though it is not desirable to enter upon it here in detail, there is a relativistic treatment of gravitational changes which shows that they are propagated at the speed of light. [6] The theory involves the existence of gravitational waves, in some ways

---

6. BONDI, H., *Gravitational Waves*, Endeavour, July, 1961, P. 124 ff.

analogous to the electro-magnetic waves set up by moving electric fields but these waves, assuming that they do exist, must be so weak that we shall probably never be able to detect them. ([7])

Despite its odd character gravity produces accelerations which cannot be distinguished from those produced in other ways. An example often used is that of an observer in a closed box who cannot know what is going on outside it. ([8]) In a region of space sufficiently far from all massive bodies the observer in a motionless box would experience the sensation of weightlessness, just as does an observer in a space capsule falling freely in the earth's gravitational field. If some space creature were to lay hold of the box and pull it upwards with a suitable constant acceleration, the observer would experience weight, just as if the box were at rest in a gravitational field. ([9]) Thus a suitable choice of acceleration does enable us to transform away a gravitational field at least in some particular direction. ([10]) This indistinguishability of gravitational and inertial forces is embodied in the Principle of Equivalence.

The next question is whether there is a geometry in which the world-lines of constant curvature representing uniform acceleration become geodesics. If this can be done it will no longer be necessary to appeal to a *force* of gravity for a geodesic is the track which a moving body will follow without any force applied, in Aristotelian terminology it is a natural motion. Einstein found a readymade technique for dealing with this question in the differential geometry of spherical space developed by Riemann nearly half a century before. What so long had been a piece of pure mathematics was to become a representation of a physical world.

We have already seen that this kind of space is finite but unbounded. The idea that the space of the universe might even be thought of as finite was shocking to very many people. The notion would not have disturbed anyone from Aristotle to Copernicus except the followers of Lucretius, but for some 300 years people had become accustomed to the idea of infinite space and the idea of finiteness came as a most upsetting surprise. The concept of the luminiferous ether had not yet been given up by all scientists and was to persist among laymen and the popular press for long af-

---

7. Ibid, P. 127 ff.
8. EINSTEIN, A. Loc. Cit. Chapter 20.
9. Weight is the measure of the force of gravity when it cannot produce an acceleration, but when it can, weight vanishes. In Newtonian mechanics it was a coincidence that inertial and gravitational mass have the same magnitude but in relativity they are necessarily identical.
10. Strictly this applies only to the field in a particular direction. We could not simulate the whole spherically symmetrical field of the earth at the one time.

terwards. Now a further shock had been administered. Einstein himself had grown up with the idea of infinite space and had been able to use it very successfully for special relativity but he found insuperable mathematical problems with infinite space in the general theory.

Basically it was the old Newtonian problem of gravitational potential. In terms of Newton's law of gravitation a uniform distribution of matter to infinity is impossible; the average density of matter at infinity must be zero. This means that from any centre the density must decrease more rapidly than $1/r^2$* where r is the distance. But no unique centre can be nominated, hence from every point of an infinite system the density must fall off faster than $1/r^2$ and the consequence of this is that the density must be everywhere zero. The only alternative is to modify the law of gravitation and, in principle, it is possible to make a modification such that a Newtonian system could exist for some extremely low average density of matter. The effect of the modification must appear only at large distances and not affect local phenomena. Although several times suggested this alternative was never acceptable and there is no observational evidence for a deviation of the kind involved from the inverse square law. At first sight a finite material system in infinite space avoids the problem but of course radiation can vanish from it into space and particles or matter with sufficient velocity can also "evaporate" from the system. In a long enough time with the consequential reduction of gravity, evaporation will in the end be complete. Nor is such a scheme compatible with our notion that on the large scale the universe is homogeneous and isotropic.

The theory of relativity does not leave the volume of finite space as an arbitrary quantity; the volume of space is determined by the average density of the matter which it contains. A completely empty universe would be infinite and open but the presence of matter brings about positive curvature. Actually any positive curvature, however slight, will result in the closure of space. When the average density of matter is very low the curvature is slight and the volume very large; this appears to be the situation in the observable universe. ([11]) While the overall curvature is determined by the material content of the universe, we have to consider that on the small scale matter occurs in sizeable lumps like stars and

---

11. Average density is determined in a volume of space in which the radius r is large compared with the distances between stars but small compared with the dimensions of the system as a whole. In any observable region of the universe the average density, though small, is clearly not zero. For **EINSTEIN'S** discussion of this subject see his *Cosmological Considerations on the General Theory of Relativity*, Dover reprint S 81, P. 177 ff.

✳ Surface of spherical shell at distance r goes as r^2.

planets. The proposition is that these modify the general curvature locally. Theoretically purely local curvature could leave space Euclidean on the large scale but distorted in the vicinity of matter. In the vicinity of masses of matter the curvature is greater than average, fading with distance into the general curvature. Thus a body near the earth and free to move is apparently attracted to it: the body simply slides along the local space curve until it reaches the surface of the earth. It works for orbits too, planets roll round tracks laid out for them in space.

In many ways this is a pleasing concept and it certainly gets rid of the awkward force of gravity but, you may say, the idea that matter can cause some modification to the properties of space is at least as indigestible as the notion of action at a distance, perhaps it is just another way to say the same thing. Before going any further let us enquire whether there is any observational evidence which can reasonably be accounted for in terms of space curvature. General relativity does offer certain predictions which can be tested. Although the observations involved are technically difficult and require refined experimental methods, they are not beyond the limits imposed by modern instruments. While not absolutely conclusive, the results of the tests appear to verify the predictions beyond reasonable doubt. The relativistic law of gravitation differs slightly from the Newtonian law which cannot account for the observed motion of the planet Mercury. The elliptical orbit of that planet slowly rotates so that its perihelion position, the position of closest approach to the sun, continually advances but it advances by 43″ of arc more in a century than the Newtonian theory predicts. ([12]) Relativity predicts a value which agrees with the observed value within the limits of experimental error. Relativity also predicts a red shift resembling the Doppler effect in the light emitted by massive bodies. Though not so dramatically verified, this prediction is at least consistent with observation and its reliability is nowadays generally accepted.

The third test is more directly connected with space curvature. If space is indeed curved, light instead of travelling in an Euclidean straight line will follow a geodesic. Similarly to the situation in special relativity it will be a geodesic along which $ds = 0$. According to the theory, this track, called a null geodesic, will be more strongly curved in the vicinity of a

---

12. At one time it was thought that this rotation was due to the perturbing influence of another planet moving inside the orbit of Mercury. This hypothetical planet was given the name Vulcan in the expectation of its discovery but there is quite certainly no such planet.

massive body. More particularly, light from a distant star passing very close to the sun on its way to us will be deflected by 1.7″ of arc when it just grazes the sun. Obviously such an observation can be made only on the brief occasions of a total eclipse of the sun. It means that a star which, for straight line propagation of light, would be just hidden behind the edge of the sun's disc would become visible by the bending of the light path. The first total eclipse after the publication of the general theory occurred in 1919, and naturally great interest was taken in the results. The position of stars which would lie in directions very close to the sun at the time of the eclipse were checked with special accuracy and, when the eclipse was over and the calculations made, it turned out that the deflection was very close to 1.7″ and Einstein's prediction was verified. ([13]) Small as the effect might be, there was no mistaking it, a spectacular piece of evidence in support of the theory.

With the encouragement given by these observations, let us look again at the conceptual situation. Neither Einstein nor anyone else has been able to propound a mechanism by which matter can cause space to be bent where it is and the ultimate cause of gravity is a mysterious as ever but quite a lot has been gained. We are in possession of a superior theory of gravitational action ([14]) which does not bring in the logically awkward force of gravity and we have acquired a mastery over the description of the observed phenomena of time, space and matter never before possible. Admittedly our attempts to visualise the behaviour of nature have proved worse than useless but, after all, what right had we to suppose that we could successfully visualise nature? All we have had to discard is a set of man-made concepts and now we adopt others better suited to the purpose in hand. Without doubt the most pernicious of those abandoned concepts were the absolutes, though long familiarity had made them seem reasonable. The criterion for adopting a concept is not whether we like it or not but whether or not it is useful. These new ideas are different from the old ones but different does not necessarily mean better or worse; it can mean simply different. When these new but disturbing notions provide a better way to describe the phenomena of nature we should adopt them until something still better appears and our feelings about them are completely irrelevant. It really does not matter that we cannot form a

13. Later observations have yielded a slightly different value and, since the eclipse of 1952, the best value seems to be 1.79″.
14. A number of different theories of gravitation have been put forward. For some examples see NORTH, J. D., Loc. Cit. P. 186 ff.

picture of curved space so long as we can use the concept in a consistent and fruitful way. Once again it is worthwhile to remind you that the term curvature is used for lack of a better word and must be taken in its technical sense as denoting a charecteristic which can be mathematically specified but which has nothing whatever to do with what we customarily think a curve looks like.

It is not possible to probe more deeply into general relativity without a much more powerful mathematical equipment than this course presupposes: in some degree it has been an act of temerity to go as far as we have. Indeed, a full understanding of the concepts and their implications is restricted to a rather small class of people who have pursued advanced mathematical studies and even for them some of the further reaches of relativity are diabolically difficult. ([15]) For cosmological purposes we must take note of one or two more consequences of the theory but they will have to be accepted without detailed explanation. ([16])

It certainly seems that general relativity ascribes to space something more than mere fitness to contain matter; variable size and curvature apparently imply properties by which it can interact with matter and energy. One of these interactions has the result that the velocity of light is not a strict universal constant but is affeced by the curved space, that is, by the gravitational fields through which it passes. The velocity is constant in a uniform field and for the weak, rather uniform fields prevailing in inter-galactic space, the variation is slight, hardly of cosmological significance. At the risk of being repetitious, it must be emphasised that the bending of light rays and perhaps theoretically the small variations in its velocity are things which can be observed. Curved space itself cannot be observed any more than the absolute space of Newtonian mechanics. It is a concept which can be usefully employed in thinking about phenomena. If no other lesson has been learned from relativity it will have been worth all the effort to realise that we deal in models of nature, not in nature itself.

It was mentioned earlier that the identity of inertial and gravitational mass is a point of some interest. In relativity these are necessarily the same; the experience of weight produced by gravity and by any other

---

15. Eddington who dug deeper than most relativists, reached a stage where in order to further his researches, he had to devise a whole new mathematical method called the "wave-tensor calculus."
16. A deeper study of relativity requires proficiency in tensor calculus, without which the development, manipulation and even the significance of the famous field equations cannot be grasped.

source of acceleration cannot be distinguished, but why do bodies have inertia anyway? It is certain that we can no longer accept the old notion that inertia comes from motion with respect to absolute space.

The most appealing proposition so far offered is that due to Ernst Mach, based on an argument similar to but not just the same as that used by Bishop Berkeley against absolute motion. Mach's principle is a statement that inertia is wholly caused by the matter of the universe and that its magnitude is determined by the mass of that matter and its distribution. ([17]) Since the vastly preponderant mass in the universe is at vast distances, the effect of local masses is not significant and we do not have to allow for the local small scale distribution. Even so it has not yet proved possible to find a mathematical formulation for Mach's principle. Einstein felt that the principle is valid and wished to give expression to it in general relativity. The outcome was that while his theory is consistent with the principle and can show that the masses of the universe have some influence on inertia, he did not succeed in showing that those masses are the whole cause of it. No other theory has succeeded better than this and it is generally counted as a virtue in a cosmological theory that it is consistent with Mach's principle.

As suggested earlier, the interval or line element ds in general relativity must contain information about the characteristics of space. Local information can be introduced into the equation in the form of coefficients attached to the space-time factors. Four dimensional space-time involves ten of these coefficients, which have the mathematical form of determinants, presented as what is technically called an "array of g's." ([18]) This form of notation is very commonly seen in books and papers but, for present purposes, the essential thing to know about them is that they contain information about the shape of ds. One might suppose that the g's would represent gradients, the rates of change of shape. Actually it is a little more complex than that, the g's are not first derivatives but second

---

17. The subjects of inertia and Mach's principle are interestingly discussed by BONDI, H., *Cosmology*, Loc. Cit. Chapter 4 and with a different line of approach by SCIAMA, D. W., Loc. Cit. Chapter 8.
18. Representing the four components of space-time by dx_1, dx_2, dx_3 and dx_4 for the sake of compactness and identifying the g's by their row and column positions as determinants (pronounced as g one, one; g two, three and so on) we get a general equation for ds^2 like—
d$s^2 = g_{11}dx^2{}_1 + g_{22}dx^2{}_2 + g_{33}dx^2{}_3 + g_{44}dx^2{}_4 + g_{12}dx_1dx_2 + g_{13}dx_1dx_3 +$
and so on for the whole set. An array of g's relates to a particular frame of reference and by setting up such an array for each observer's frame it follows that they will all find the same value for the ds of the same pair of events, just as in special relativity.

derivatives representing the gradients of gradients, for it proves necessary to show how the rate of change of shape changes.

It would be reasonable to think that electric and magnetic phenomena could also be handled by a theory involving space curvatures analogous to that used for gravitation and this can certainly be done; consequently much effort has been put into attempts to produce a unified theory which could cope with electric and magnetic fields as well as the gravitational field but, so far, without success and it does not seem very likely that it will be achieved in the immediate future. In recent years some developments of microphysics seem to have moved further away from any possibility of a unified theory. Thus general relativity remains largely a theory of gravitation.

Special relativity is mostly concerned with laws of motion and so, in a broad sense, is general relativity, whereas one may say that cosmology is concerned with actual motions. The distinction is not as clear as it looks: it certainly does not have the strength which it has in the Newtonian system. Indeed some cosmologists do not make the distinction at all and it has really been mainly used here as a matter of convenience rather than of principle. All modern cosmological theories accept the validity of special relativity for local affairs and many, though by no means all, base their models on general relativity. Those which do not have still been largely influenced by it, for a mode of thought has been introduced which is far too useful to be neglected even if a different approach appears in the model. The next task is to see what kinds of models of the real world can be built around the concepts of relativity.

*The Stars are setting and the Caravan
Starts for the Dawn of nothing.*

CHAPTER 11

THE FIRST RELATIVISTIC COSMOLOGIES

The idea that the space of the universe around us might be closed inevitably called for the development of finite cosmological models based on relativity and it is fitting that the first attempt should have been made by Einstein himself. You will naturally wish to know what this first relativistic venture looks like and the answer is: peculiar.

Starting with an array of descriptive adjectives, Einstein's universe is finite, unbounded, static, homogeneous and isotropic. Four out of these five sound thoroughly Newtonian and one may be pardoned for wondering whether this is going to be the first of the relativistic or the last of the classical models. It is perhaps fairest to say that it is a hybrid transitional model. Whatever view is taken, the model presents some serious conceptual problems. ([1]) The feature of being static, which was soon to be the death of the model when the expansion of the universe was discovered, strongly emphasises its hybrid character. Its static uniform distribution of matter constitutes a unique frame of reference, a universal standard of rest, in just the same way as did Newton's absolute space and the luminiferous ether. There could hardly be anything more incompatible with the spirit of relativity.

Further, the static quality of the model was obtained only by appeal to an *ad hoc* hypothesis which, at the best, could give it no more than conditional stability. This rather sweeping denunciation should not be taken as suggesting that the founder of relativity could not put it to work. Einstein made the model static simply because every astronomer could assure him from the observations then current that there were no systematic motions on the large scale. As to the *ad hoc* hypothesis, Einstein was fully aware of its character and, having been forced into it at the start, got rid of it as soon as possible though others took it up in a modified form with considerable enthusiasm. After all a beginning had

---

1. To be truthful, quite some conceptual problems have been passed over in the previous two chapters, not in the hope that they will have gone unnoticed, but because it was simply not practical to cope with them all in a course of this nature.

to be made somewhere and it was Einstein who ventured on to the unknown ground.

To appreciate the significance of the model for future thought, we must enquire more specifically into some of its properties. It is fairly obvious that a finite volume of space cannot be expected to accommodate an infinite content of matter and it is reasonable to suppose that the total mass of matter will be a definite finite quantity. The value of R which determines the size of space was found by Einstein to be related to the mass M thus:

$$R = \frac{\gamma M}{\frac{1}{2}\pi c^2}$$

where γ is the universal constant of gravitation. The mass of matter and the size of space can also be related through the average density of matter ϱ:

$$M = 2\pi^2 R^3 \varrho$$

as the volume of the spherical space is $2\pi^2 R^3$. It is to be seen that R increases with M but, for a given M decreases with ϱ so that space is always just big enough to hold the matter of the universe. Einstein took a value for ϱ around $10^{-29} \mathrm{g/cm^3}$ which is in fair agreement with observed values. [2] If all the matter of the universe was squeezed into a state of high density the volume of space would become very small, a situation which arises at the earliest stage in Lemaître's model. The statement made very early in this course, that space is where matter is, still applies.

Einstein was able to arrive at definite values of R and M through his *ad hoc* hypothesis which took the form of a universal constant λ, called the cosmological or cosmical constant. λ appears in his law of gravitation as a factor having no significance on a small scale such as the dimensions of the solar system, but becoming important at large distances comparable with R. [3] It will be recalled that Einstein was anxious to incorporate Mach's principle into the theory and he felt that λ did this by ensuring that there was a solution to his field equations for a small but definite value of ϱ and he believed that there was no solution for zero density. In other words, the factors in the equations representing inertia would be determined by the mass of matter in the universe and its uniform distribution.

From Einstein's point of view λ was essential to provide for the static quality of the model by modifying the law of gravitation in a way

2. Cf BONDI, H., *Cosmology*, Loc. Cit. P. 45 for observed values.
3. NORTH, J. D., Loc. Cit. P. 70 f.

operative at great distances and so does what a suitable modification of the inverse square law would do for an infinite Newtonian universe. Mathematically λ comes to have the value given by:

$$\lambda = \frac{\varkappa\varrho}{2} = \frac{1}{R^2}$$

where $\varkappa$ is a factor containing information about gravitation. You may like to consider it as a translator, putting the Newtonian universal constant of gravitation into relativistic form. [4] Since R is large it follows that λ is a very small quantity and thus insignificant on the local scale. At this stage it is appropriate to quote Einstein's own thoughts on λ [5]

"Thus the newly introduced universal constant λ defines both the mean density of distribution which can remain in equilibrium and also the radius R and the volume $2\pi^2 R^3$ of spherical space. The total mass M of the universe, according to our view, is finite, and is in fact

$$M = \varrho 2\pi^2 R^3 = 4\pi^2 \frac{R}{\varkappa} = \pi^2 V\left(\frac{32}{R^2\varrho}\right)."$$

Now we have fairly definite values of λ, R, ϱ, and M, all interlocked and no independent criteria for most of them. In the whole basis of this set of quantities, only ϱ had a direct grounding in observation and the relationship of the four factors is essentially one of mathematical theory. So far as practical observations were concerned only ϱ had a reasonable prospect of verification, R was too large, λ too small and M was unattainable. If we feel uncomfortable about this situation, we are not alone. [6]

It is apparent that λ was neatly proportioned to provide a balance for gravity, keeping the model static for a definite value of ϱ. As Einstein clearly saw, this is an *ad hoc* contrivance and appeal to observation will not help, for that is the tautological procedure of calling on the result to justify the explanation made to account for it. [7] There is a very remarkable physical implication brought up by λ; it implies a force of repulsion to balance gravitation on the large scale and this force, if real, is unique

4. γ has the value 6.670×10^{-8} dynes/g/cm^2. $\varkappa$ is $\dfrac{8\pi\gamma}{c^2}$ and has the value 1.864×10^{-27} cm/g^{-1}.
5. EINSTEIN, A., Dover Reprint, Loc. Cit. P. 187.
6. NORTH, J. D. Loc. Cit. P. 83.
7. Compare the anecdote of the man who was asked why he was scattering scraps of imaginary paper around him and replied that it was to keep away elephants. When it was objected that there were no elephants in the vicinity, he claimed that this demonstrated the efficacy of his method.

in that it increases with distance. No observable force behaves like that. ([8]) None the less it leads to an intriguing speculation. Suppose we could have a model with slightly different proportions; then maybe the cosmic repulsion might not merely balance gravity but actually overcome it. In that case every point of observation in the universe would be a centre of repulsion from which very distant objects would be seen to recede with ever increasing velocity. Almost before anyone had the time to contemplate this thought at leisure, something astonishingly like it turned up.

Willem de Sitter did not intend to make a model with a feature of recession, for the astronomical position was the same in 1917 as it had been a couple of years before when Einstein published his theory; and de Sitter was himself an astronomer. What he set out to do was to find some other solution to the field equations than the one which Einstein thought was unique. When de Sitter found his solution it was for zero density of matter. This did not mean a very small density; it meant none at all. It might seem at first sight that an empty universe, containing no matter at all, would not contain any interest either but this was not the case for the theoretical interest is considerable. The model is static because, as de Sitter pointed out, perhaps with a touch of sly humour, there is nothing in it to change. There is something disturbing to our intuitions in suggesting that empty space could have an intrinsic geometry but it seems to be the non-Euclidean aspect that is objectionable, for no one complains very loudly when we talk about empty space in Euclidean terms. Perhaps it will lend an air of respectability to the proposition if we extend our original definition of space to read: space is where things are or could be. The space-time of de Sitter's universe turned out not to be spherical and closed, but open and hyperbolic.

One thing is clear; regardless of any hypothetical properties of space, the fact that a solution is possible for zero density shows that Einstein's model did not succeed in incorporating Mach's principle. Those who hold with the principle, including Einstein, felt that this was a deficiency in general relativity itself, while those who were opposed to the principle took this as evidence of its unreality. It seems that a test particle of negligible mass injected into a de Sitter world would have almost the same inertia as in the massive Einstein world. On this basis it could be said that de Sitter was not dealing with the void, but with the geometry of an inertial field.

---

8. It is hardly surprising that Newtonian thinkers were generally unwilling to accept any tinkering with the inverse square law such as had been suggested by Seeliger before the days of relativity.

Out of this shadowy world came some ideas of deep theoretical interest. Remember that it was a world utterly divorced from observation, quite overtly a man-made mathematical construction. Though in its strict form it could never become a model of the real world, de Sitter's universe had a strangely prophetic quality, giving astronomers a strong hint about what to look for in the depths of space. One aspect of the prophecy was in what soon became known as the de Sitter Effect.

Mathematical considerations led to the proposition that if we could put into this empty world some sources of light having negligible mass, the light from very distant ones would be shifted towards the red end of the spectrum. So far as an observer is concerned, either the frequency of the light decreases in transit, or at great distances it is generated with a lower frequency than locally. The latter view was expressed in the statement that time proceeds more slowly at great distances. [9] In either case, a hypothetical observer finds he has an observational horizon at a finite distance; that is the distance from which the frequency of light is reduced to zero or at which time appears to stand still. No knowledge can be gained of events beyond the horizon, which defines a spherically symmetrical observable region of the universe for each observer. Of course, an observer close to our horizon would see us as close to his and he would be able to see events invisible to us.

Before this there had not appeared a theoretical reason why space should not be penetrated as far as we please with the aid of telescopes of sufficient light grasp; the limitations were supposed to be technological rather than theoretical. In the de Sitter case we are faced with the principle that zero frequency is the limiting case where radiation ceases to exist, when its energy has fallen to zero. Observational horizons in one form or another have been a feature of subsequent cosmological models.

The reddening of the light in the de Sitter model resembles the Doppler effect, already well known to result from the motion of a light source relative to the observer. The other prophetic feature in de Sitter's model [10] was that if a number of particles of negligible mass could somehow be put into the model, they would not remain stationary but would recede from each other with ever increasing velocity. This great contrast with the static situation in Einstein's model gave rise to the expressive

---

9. It is really much more complicated than that and a great deal hinges on the definition of the distance used, i.e., proper or coordinate distance. For an account in considerable detail see NORTH J. D., Loc Cit. P. 92 ff.

10. EDDINGTON, A. S., *The Expanding Universe*, Pelican reprint, A 70, P. 1 f. (My copy is of 1940 vintage.)

comment that the Einstein universe contains matter but no motion while the de Sitter universe contains motion but no matter. ([11])

If the receding particles were light sources they would exhibit a Doppler effect into which it would be possible to absorb the de Sitter effect as a minor factor. Eddington seems to have been one of the first, if not the first, to envisage the possibility of using de Sitter's model in a modified form, including a very small density of matter, as one in which there would be a genuine recession. There were two or three galaxies known to have a marked red shift and Eddington came out strongly for real recession.

Within a short time after this essentially theoretical exercise evidence began to appear for a red shift in the light of very distant bodies. Hubble, Humason and Slipher were finding red shifts in the light of the spiral nebulae, which seemed to be greater in proportion to their distances. By 1922 Slipher was able to provide Eddington with a list of some forty nebulae with reddened light and so, presumably, with motions of recession. ([12]) Astronomers were now taking as much interest in the de Sitter model as the mathematicians for, if as seemed likely, there are general motions of recession only something like de Sitter's model offered a prospect of accommodating it. Obviously a completely empty model would not do, but if there were solutions for the equations for intermediate values of density, between the empty de Sitter model and the "full" Einstein one, it might be possible to arrive at one containing both matter and motion. There were no well established observational values for the average density of matter at that time and a compromise close to, but not quite empty like the original de Sitter model, might be a fair representation of the real universe. Such solutions were found, notably by Friedmann, Lemaître and Robertson, Between them, de Sitter and Eddington, the one almost in spite of what he set out to do and the other with deliberate intent, set a new pattern. Since then all cosmological models have been non-static and must account for the observed red shift, or at least for the constructor's interpretation of it.

The expansion of the universe is an essential feature of modern cosmology so we will be well advised to learn something about it before going any further. There is quite some preliminary ground to be cleared. First

---

11. I do not know who first said this but it has been quoted by Bondi, (1952), Hubble, (1936) and Eddington, (1932).
12. Actually out of 40 galaxies, 36 showed a red shift and only 4 a violet shift. Subsequently correcting factors eliminated this minority. Students should read HUBBLE, E., *The Realm of the Nebulae*, Dover S 455, 1958

of all the objects with which we are concerned in discussing the expansion
are the galaxies, typified by the spiral nebulae on which Kant built his
theoretical model and which so greatly interested Herschel and Rosse.
You will recall that the spectroscopists of the 19th century had contrived
to rob the spiral nebulae of their extragalactic status, even though their
total lack of parallax or proper motion implied that they were very far off.
Early in this century a means had been found to estimate their distances.
The secret lay in the discovery that certain very bright variable stars
have a definite relationship between their period and their absolute brightness. ([13]) The distances of some of these Cepheid variables, so called from
δ Cephei the type star of this class, were approximately known. Some of
these stars could be identified on photographs of the nearest galaxies like
Andromeda and, however rough the measures might be, it soon became
certain that these objects are at distances large compared with the dimensions of the Milky Way. More distant galaxies do not permit the resolution
of individual stars, but some idea can be gained by the comparison of
their brightness with those nearer to us. ([14]) There is an assumption here
that the absolute magnitudes of the galaxies are more or less the same, at
least for those of similar structure. Though far from accurate, if it is
recognised as a working hypothesis, to be modified as new knowledge is
gained, it will be a fair starting point.

In broad terms, the expansion of the universe means that the distances
between the basic cosmological units, the galaxies, is increasing. It is all
too easy to fall into the trap of thinking in customary terms that expansion
means expansion into surrounding regions. Behind this the old notion of
absolute space lurks in the background; when something expands it
occupies more of that space. This idea cannot work for the universe as
a whole; it already occupies all space, and we have become accustomed
to the idea that space can adjust itself to its material content. If the system
of matter expands, the density becomes less because there is less matter
per unit volume and space becomes bigger, simply because space is where
matter is.

There is no suggestion that the expansion of the universe relates to
things on the small scale; it does not mean, for instance, that the earth
expands, nor the solar system, nor the galaxy itself, perhaps not even the

13. LEAVITT, H., *The Discovery of the Period-Magnitude Relation*, Reprinted in
    Shapley, S.B.A., Loc. Cit. P. 186 ff.
14. Also nowadays with radio telescopes, the "brightness" is measured at the longer
    wavelengths of radio waves.

clusters of galaxies. These structures are gravitationally bound; the scale of size is such that in these cases gravity predominates locally over anything which might cause dispersal. Thus the expansion of the universe has nothing to do with Reichenbach's proposition, "the nocturnal doubling statement" in which the question is whether we could know about it if, during the night, the dimensions of everything in the universe, atoms and all, were suddenly doubled. ([15]) In principle we can know about the expansion of the universe because our proposition is that the distances between certain objects in it are increasing.

To make an analogy in terms of a finite but unbounded space of two dimensions, let us consider the surface of a sphere which, for the purposes of illustration, can be the surface of a rubber balloon slightly inflated. Now with a little care this surface can be marked with small dots evenly spaced all over it; we might preferably do this by glueing a grain of sand at the position of each dot, then like galaxies, these would not expand in themselves when the balloon is further inflated. The consequence of further inflation is that the grains become farther apart. An observer on each grain would see all the other grains moving away from him. For a particular rate of inflation he would see the first grain on his right moving away at a certain speed and the second in the same direction moving away twice as fast and so on until, halfway round his world, the most distant grain to be seen would have the highest velocity of recession.

Even in this simple model things are not quite as simple as they seem. Something rather disturbing happens to an observer's concept of distance. He will have started off with a quite adequate concept of distance which materialised as some kind of a yardstick. Special relativity will have taught him to be careful how he uses it, but the Lorentz transformation will enable him to make suitable corrections to his observations and a rather large distance like the diameter of his sand grain galaxy can be clearly stated in terms of his yardstick. The other grains can only be observed by means of light which reaches him a long time after it was emitted and is reddened by the Doppler effect. This reddening, which can be accurately measured, ought to tell him about the velocity of recession which increases with distance, but that distance is increasing while the light traverses it. If the expansion continues long enough, our observer will no longer be able to see halfway round his world, for the velocity of recession will

---

15. There is an implication here that despite the doubling there remains something left over from the previous state of the universe to act as a reference, otherwise the statement is meaningless.

become so great that the light from very distant sources will not be able
to overcome the increasing distance and so will never reach him. Thus
a light horizon appears and our observer reports to his director of research
that not only do distant bodies recede with velocities approaching c, they
can exceed it and there must exist galaxies beyond this light horizon which
he could have observed had he lived a hundred million or so years earlier
but now it is too late.

This apparent contradiction of what we have learned about material
bodies never being able to exceed c ([16]) comes about because our observer
is trying to work with two quite different concepts of distance. The
velocity of light has been defined in terms of local or proper distance and,
in these terms, c is constant or near enough to constant, but these
distances a long way off are quite another proposition. Instead of proper
distance we can consider coordinate distance, where the position of an
object can be specified in terms of a mesh system such as we discussed in
Chapter 10. On the small scale of local affairs, coordinate and proper
distance can be neatly tied together. But, when we come to the large scale,
it is clear that the relative positions of galaxies do not change, in the sense
that their order of succession in the mesh remains unaltered. Really what
we are saying amounts to a statement that the arrangement of the galaxies
in space constitutes the mesh in which, from the viewpoint of proper
distance, the size of a mesh unit depends on where it is. In terms of a
mesh determined by the galaxies, that is coordinate distance, the velocity
of light is certainly not constant. If at this stage readers are becoming a
little puzzled about what distance really means, more good than harm
will have been done. Out of this confusion should come the idea that there
is nothing absolute about distance and the measurements we get depend
on how we go about getting them. There is in fact a relative quality about
distance: however we define it, we can get a meaningful answer to measurements providing we are consistent. Whether or not the answer is in the
same terms as our rather intuitively based idea of proper distance is not
important and, with due care, conversions can be made. It may occur to
the reader that a purpose of metrical formulae is to make possible the
transformation between proper distance and other kinds of distance, for
instance coordinate distance and this indeed they do. In passing, we may

16. Special relativity talks about the velocities produced by accelerating forces.
This is quite different from velocities produced by a change in dimensions of
the system.

take note that the idea of proper distance is not quite as simple as it seems. While the mathematical definition is plain enough in terms of some standard unit of length like a metre or the wavelength of some particular spectral line, ([17]) there remain some conceptual problems in using it beyond the possibility of direct application.

In a static spatial system the positions of neighbouring points can, as we have already seen, be determined in terms of a set of coordinates, say r, θ and φ. These can be expressed in terms of proper distance providing we know the law according to which the mesh system grows with the distance. In an expanding universe where the mesh is not static, we have to know how it grows with time. In a metrical formula for an expanding universe it is thus necessary to be able to express the relation between proper and coordinate distances at any epoch, so in it we find $R(t)$, the scale factor at epoch t, and k which specifies the kind of curvature. An example is one of the forms of the Robertson-Walker line element for relativistic models which looks like this:

$$ds^2 = dt^2 - R^2(t)\,(dr^2 + r^2 d\theta^2 + r^2\sin^2\theta\, d\varphi^2) / (1 + \tfrac{1}{4}\,kr^2)^2$$

for positive curvature $k = +1$ and for negative curvature -1, while the value 0 means the space-time is flat.

The distances of the nearest galaxies, determined by observations of Cepheid variables and supplemented by the occasional nova star of much greater brightness, despite their inferential character can still be considered to be local or proper distances. Beyond this range all distances are necessarily luminosity distances. To give luminosity distance a significance by which it can, in some sense, be meaningfully compared with local distance, it must be clearly defined. It will of course be an operational definition. ([18]) Given the rate at which light is emitted by an object of observation, that is its luminosity, distance is inferred from the comparison of the emitted and received intensities of light. ([19]) Though it sounds simple when expressed in this way, luminosity distance involves some quite complex factors. A fairly obvious one is that allowance must be made for the loss of light absorbed by obscuring matter in our own galaxy, essentially dust. Dust does not just absorb light, it absorbs it selectively, the violet end of the spectrum being more strongly affected. The phenomenon

---

17. I cannot resist reference to the apocryphal character who said that the standard horsepower is the distance between two marks on the back of a platinum horse kept at Paris.
18. NORTH, J. D., Loc. Cit. Chap. 15.
19. BONDI, H., Loc. Cit. P. 68. Also McVITTIE, G. C., *Fact and Theory in Cosmology*, Eyre & Spotteswoode, 1961, P. 89.

is well known here on earth where the sun is seen to be redder through smoke or dust haze. Nowadays the obscuration in various directions is fairly well known. Two other still more local effects must be allowed for; the transmission of the atmosphere and the colour sensitivity of photographic plates and these too are now well known.

The Doppler shift itself must now come under scrutiny. The red-shift of a body receding in a Newtonian universe is very simply expressed in terms of wavelength. Taking some particular spectral line, its wavelength as measured in a laboratory here is λ. In the spectrum of a body with red-shift, that same line has a wavelength $\lambda + d\lambda$. The Doppler shift has no colour discrimination so the displacement of all the spectral lines on the photographic plate is $d\lambda/\lambda$, often conveniently expressed as $1 + \delta$. It then follows that the velocity of recession V giving rise to the red shift has the value $V = c\delta$.

If an examination of the spectrum gives the same δ for all lines, that is the shift is always proportional to the wavelength, it is reasonably safe to assume that the cause of the red shift is a velocity of recession as almost every other known cause of reddening shows colour discrimination. This classical Doppler formula can always be used for small red shifts where the velocities, like those of stars in our own galaxy and of nearby galaxies are small compared with c which implies that the distances are small.

At great distances and in an expanding universe the interpretation of the Doppler effect must take $R(t)$ into account. At the source the time difference between the emission of two successive waves of some particular wavelength may be written dt, from $(t + dt)$. At 0, the observing position, the waves will be received some time later with a time difference between successive waves dt_0, from $(t_0 + dt_0)$ and it can be shown that these are related so that: [20]

$$\frac{dt}{R(t)} = \frac{dt_0\delta}{R(t_0)}$$

where $R(t)$ and $R(t_0)$ are the values of R at the times emission and reception, whence $1 + \delta = R(t_0)/R(t)$. Of course $R(t)$ is known only when a specific model of the universe is being used. As the Doppler effect alters the energy content of the light as received, it must be included in the formulation of luminosity distance. Actually it comes in twice, [21] so in the final expression we find the Doppler factor $(1 + \delta)$ squared. It

20. McVITTIE, G. C., *Fact and Theory in Cosmology*, Eyre and Spottiswoode, 1961, P. 109.
21. BONDI, H., Loc. Cit. P. 108.

obviously has to come in once because of the altered wavelengths of the light. It is brought in the second time since not only is the energy of the received photons reduced, but the number of them received in a given time is reduced in the same ratio.

The light of the galaxy under observation is emitted in all directions and of course we receive only a tiny fraction emitted over a very small solid angle. In Euclidean space the proportion can be found quite simply from the inverse square law, based on the propagation of light in straight lines. This geometrical factor, $\dfrac{\text{absolute brightness}}{4\pi \times \text{apparent brightness}}$ and the local corrections mentioned above are all that is necessary to give luminosity distance in a static Newtonian universe. As curvature and expansion have to be taken into account $R(t)$ must be used to derive the equivalent propagation law and the change of distance during the transit of the light. For luminosity distances on the local scale the propagation is near enough to Euclidean and R does not change appreciably during the short time involved so the classical formula can safely be used.

When all these factors are taken into account we arrive at the formal expression for luminosity distance:

$$\sqrt{\left(\frac{L}{4\pi L_0\,(1\,+\,\delta)^2}\right)}$$

where L and L_0 are the emitted and received intensities measured in proper units at the source and at the observer.

Luminosity distance enables us to arrange the observed galaxies in order of distance which amounts to assigning them coordinate distances. As there is something more or less resembling a local distance for nearby galaxies, it is possible to extrapolate local distance to the far off galaxies, providing one does not object to extremely large limits of error in numerical values. An example of the kind of surprise that can occur in doing this, is the revision some years ago by Baade of the distances of the Cepheids. He found that they are twice as far away than had previously been estimated. Consequently, the local distance of the nearby galaxies had to be doubled and the scale of luminosity distance was doubled in terms of local distance.

The early determinations of red shift were limited by the technical facilities available to comparatively nearby galaxies but the development of more advanced instruments has made it possible to survey the galaxies out to distances where the apparent magnitude is 21m. Recently radio telescopes have been able to detect objects which seem to be even further off. The possibility, no longer purely theoretical, of putting telescopes outside the earth's atmosphere on space platforms or on the moon promises a further extension of the useful range of optical telescopes both in distance and in the gamut of radiation detectable.

What the astronomers measure is the red shift of a galaxy and its apparent magnitude, so that the distances involved are luminosity distances. In these terms a relation was found, now famous as Hubble's Law, stating that the distance is directly proportional to the red shift. At first the red shifts were rather small and it was thought that luminosity distance was firmly tied to local distance by the Cepheid variable star procedure.

It now appears that Hubble's Law holds as far out as the measures can be reliably made but, of course, this does not mean that it may not require modification at still greater ranges. Nowadays distances like five or six thousand million light years are often quoted, but it must be remembered that these are enormous extrapolations of local distance and should not be taken quite literally.

From the red shift, if it is a real Doppler effect, can be inferred a velocity which must also be directly proportional to distance. In recent years such large red shifts have been found that the velocities of recession must be about half the velocity of light. Even in the days of small red shifts alternative explanations were sought and these very large velocities still make some scientists very doubtful of their reality. Though a number of alternative explanations have been offered, not one of them has proved widely acceptable. ([22]) All things considered, it seems best to take the red shifts as genuine Doppler phenomena, despite the enormous velocities.

Whatever precise significance may be given to distance, the velocity/ distance relation should be linear according to Hubble's Law. For small velocities this may be written as $v = rk$ where v is the velocity of recession, r is the distance and k is a constant. Here v and r are small compared with c and R respectively. This is legitimate when r and v are in terms of local distance. The constant k is also dependent on r and must take the form of a rate as it shows how v changes per unit distance. In the conventional representation of the velocity distance relation, we put:

$$v = r\frac{1}{T}$$

where the distance is expressed in megaparsecs (mpc) and v is in km/sec∕mpc. The early value for the velocity/distance relation was 540 km/sec/ mpc, but the drastic revision of the distance scale by Baade reduced this

---

22. NORTH, J. D., Loc. Cit. P. 229 f.

by a large factor and a recent value given by Sandage is 75 km/sec/mpc. $\frac{1}{T}$ is called Hubble's constant, though in terms of very large shifts and very great luminosity distances it is not necessarily constant. As McVittie and Bonnor have proposed, it would be more appropriately called Hubble's parameter or the Hubble factor. Mathematically, the reciprocal of the rate $\frac{1}{T}$ looks like a time and has been generally interpreted as such under the name of the "age of the Universe." For the old value of the Hubble factor T is 1.8×10^9 years and for Sandage's value, 1.3×10^{10} years.

The idea that T can represent the age of the universe comes about in this way; if we extrapolate backwards in terms of the Hubble factor, then a long time ago, the galaxies which are now observed to have mutually receded to vast distances must have been very much closer together and, if we naïvely extrapolate to the limit, there must have been a time when they were all together at the same point. For obvious reasons a strictly point origin for the universe is not satisfactory and it is usual to avoid this extreme by appeal to some rather dense congregation of matter having a finite size which represents the beginning of things.

Tomorrow? Why Tomorrow I may be
Myself with Yesterday's seven thousand Years.

CHAPTER 12

SOME RELATIVISTIC EXPANDING UNIVERSES

Although the de Sitter model was very prophetic in suggesting that a red shift and an expansion might be observed, it did not prove suitable in the long run to represent the real world where the red shift turned out to be rather bigger than the de Sitter effect. A number of cosmologists set about the task of constructing expanding models and among them Sir Arthur Eddington soon became prominent, both scientifically and with the general public. His charmingly written little book, "The Expanding Universe" of 1932, was a highly successful work of popular science. It was for his book "Mathematical Theory of Relativity" published in 1924 that Slipher had made available in 1922 the then unpublished list of nearly 40 receding galaxies, so Eddington had some very good information as a starting point.

Eddington was far too sophisticated a thinker to choose a point source for his universe and he chose for the initial condition a static Einstein type model in which the distribution of matter was uniform on both the large and small scales, occupying a finite volume of space but for which no age could be specified. For Eddington age or measurable time had to be operationally defined in terms of the succession of events and in the true Einstein model there were no events. Thus Eddington did not have to specify a date for the creation of the primordial matter; to do so would have been meaningless. It resembled in this way the Chaos of the Greek thinkers so long ago. ([1]) While the first event in Chaos was scientifically causeless because supernatural, Eddington's first event was equally causeless because it happened by pure chance. In an Einstein model the cosmic repulsion and gravitational attraction are in exact balance but the slightest departure from uniformity in the distribution of matter will upset it in the end. The density of matter in this static stage was such that there would be roughly one hydrogen atom per litre, about a four-inch cube of space. Sooner or later, after some quite indefinable period, here and

---

1. EDDINGTON, SIR A. S., Loc. Cit. P. 59.

there small random movements would result in there being a few atoms together in such a small volume and their joint gravitational field would then bring in others. Consequently the balance would be upset and in various places, a slow but accelerating process of condensation would set in, resulting ultimately in the formation of the galaxies with all their content of stars and gaseous nebulae.

You may be interested to ask where the heavier elements come from, as Eddington started with hydrogen. He did not go into this question in detail in his cosmology but Eddington was one of the founders of the theory of stars which explains how, deep in their interiors where the temperatures and pressures are enormous, thermo-nuclear reactions can fabricate heavier elements out of hydrogen. As will be seen later in this chapter, there are some problems in trying to account for the heavy elements without the aid of stars.

The collection of matter into relatively dense local aggregations means that on the local scale gravity preponderates but on the largest scale, cosmic repulsion must win. Thus expansion starts and now that backward extrapolated "age of the universe" begins to have meaning. It follows from the finite velocity of light that our part of the universe is the oldest we can know about, for wherever we look to any distance we see it some time, maybe a very long time, ago. While our local galaxy may possibly be younger than some, it certainly cannot be older than the universe in general and so the age of our galaxy, the stars in it and even the age of our earth combine to set a lower limit to the age of the universe. Now there is some evidence, geological and physical, the latter derived from the decay of radioactive atoms, that the earth is all of three thousand million years old as a solid body. How long before that the material of the earth may have existed as small particles or as gas there is no knowing. The sun, and hence the galaxy, must be at least as old as the earth and so we have a minimum age for the universe, set by the determinable age of the earth.

The early value of T was a great problem to cosmologists [2] and, had Eddington lived to see the revision of the distance and time scales, it would have involved a very drastic alteration to his model, indeed until the revision, the time scale plagued every model based on general relativity and some remarkable expedients were adopted to get around it. Eddington had to suggest that in the early stages of expansion the rate was much

---

2. A longer time scale had been proposed by Jeans, based on stellar dynamics but, unfortunately for the cosmologists, Jean's position proved untenable.

lower as the cosmic repulsion, though gaining the upper hand, could initially produce only a small velocity of recession, which grew as increasing distance weakened the restraining power of gravity. With the passage of time Eddington's universe expands towards the empty de Sitter state which it approaches asymptotically.

Noting that there are several possible values of the Hubble factor, depending on the corrections applied to the observations, Eddington took the velocity/distance relation to be 528 km/sec/mpc which entailed fairly specific values for other quantities in his universe. Before the expansion began, the static stage had a "radius" of 1068 million light-years and a density close to 10^{-27} g/cm^3. This leads to a total mass in the universe of 2.14 x 10^{55}g. As the rest mass of the proton was known, Eddington was able to find a very large number N representing the number of massive particles in the universe. At the time he wrote "The Expanding Universe," the neutron had just been discovered and it was usual then to regard it as a proton and an electron tightly bound together. So Eddington's N, which he described as the number of protons in the universe could include the neutrons. The rest mass of the electrons he could afford to neglect, as an equal number of them would have less than 1/1800 of the mass of the protons. That the numbers are about equal follows from the general electrical neutrality of matter. Evaluated in this way N comes out as about 10^{79}. Several other large numbers appear which have simple ratios to N, like $\sqrt{N}$ (3). In this class of numbers we find the ratio of electrical to gravitational force between a proton and electron and the ratio of the "radius" of an electron to the "radius" of the universe. Knowing that R and the mass of matter in the universe are closely related, it would be remarkable if Eddington had not produced an intriguing relation between R and N. Actually two values of R, both very large numbers, have to be considered, the initial Einstein R and the final de Sitter R. It appeared that the universe was well on its way to the de Sitter stage,* so the de Sitter R could well be used, with the advantage that, representing empty space, it is constant. This value of R is 3 times the Einstein R and for it λ becomes $3/R^2$ instead of $1/R^2$, with the numerical value 9.8 x 10^{-55} cm^2.

Now R is the scale factor which ties up length in the macrocosm and by bringing N into the picture Eddington obtains the quantity $R/\sqrt{N}$,

3. EDDINGTON, SIR A. S., Loc. Cit. Chapter 4.
The construction of some of these large numbers is neatly displayed by BONDI, H. Loc. Cit. Chapter 7.

* Prob. means in terms of emptiness, not that it becomes negatively curved.

which he can justify mathematically. When we look at this quantity it is seen that it represents a small length, for R is around 10^{27} cm and $\sqrt{N}$ is 10^{40} near enough, giving the result 10^{-13} cm which is the quantity that used to be called the "radius" of the electron in the days when the electron was pictured as a sort of billiard ball of energy. Today the image of the electron is, to put it mildly, more blurred than that, but a distance of the order 10^{-13} cm is significant in a newer sense as the range of nuclear forces, the size of an atomic nucleus being in the order of 10^{-12} cm. Perhaps there is no smaller significant spatial distance than 10^{-13} cm, a space quantum.

. Our references to Eddington's work in this field are necessarily superficial because of the technical difficulty of the subject, but it can be appreciated that there are certain very large numbers with relations that come out very simple in the end. It seemed to Eddington that these numbers have these values because they are necessary values, containing fundamental truths about the universe. The ultimate simplicity of their relations, he took to be clear evidence of their correctness. This sort of aesthetic criterion often appears in the mathematics of science; both symmetry and simplicity give a comfortable feeling of being on a right track, though not necessarily the one on which the venture began.

It has since been learned that these particular numerical values are not essential features of the universe. Numerical coincidences can still be set up and they are, nowadays, regarded simply as coincidences arising out of the way we define and measure quantities. Even in finite models, N no longer looks like an immutable quantity so Eddington's number lore has gone into discard. None the less he was, in a deep way, right: there must be a fundamental set of relations between the atom and the universe. Some day we shall have to find it, but it will probably take a while yet. Meantime, a fair verdict is not that Eddington was "off the beam," but that there was as yet no beam and he was trying to show where a beam might be. The attempt was premature but it was beautiful.

Though Eddington used the cosmical constant in much the same form as did Einstein and he had to postulate a cosmic repulsion, the concept of λ became rather different in his work. He saw it as an aspect of the links between the physics of the very large and the very small, which, among other things, related the size of the electron to the size of the universe. Thus he commented that the theory of the expanding universe might also be considered as the theory of the shrinking atom. However, Eddington was probably not serious about this for he was quite clear in his mind about the reality of the expansion. While the idea of the shrinking atom can be shown to be mathematically equivalent, one could nominate physical objections to it and cosmologists would expect to find it inconsistent with ideas like Weyl's postulate.

In Eddington's model the recession clearly involves a theoretical insurmountable horizon as it expands from the Einstein state towards the de Sitter state. In principle it is possible for a ray of light to go right around an Einstein universe, providing that it is quite uniform and transparent. The size of the static stage used by Eddington involved a transit time for the round trip approaching 7,000 million years. When the expansion got under way a stage was reached when distance was increasing too rapidly for light to go right round, though it would still be theoretically possible to see all the galaxies so long as light could get halfway round. Once distances had grown so much that light could no longer get halfway round, some galaxies became unobservable and the proportion of them increases with time until, at some stage, no external galaxy at all would remain within the range of observation. For observers here the observable universe would then consist of our own galaxy alone in otherwise empty space. In many ways this isolation would make little difference to our way of life. There would be somewhat less light in the night sky, but for naked eye observations it would not be conspicuous as most of the luminous objects to be seen are in this galaxy. At least one interesting experiment could be made to determine the validity of Mach's principle; if in these circumstances the inertia of bodies on the earth is greater in the plane of the galaxy than at right angles to it, Mach's principle would gain strong support.

Now that we have become accustomed to thinking in terms of finite closed space and the expansion, the existence of an horizon such as Eddington specified, seems quite a simple idea. It is time a little confusion was brought in on the old principle of *lucus a non lucendo*. In an earlier chapter it was mentioned that, at fairly regular intervals on the average, there is a supernova outburst in a galaxy, resulting in what may fairly be described as a bright flash of light. Let us for the sake of discussion, suppose that we can continue to observe some distant galaxy during vast periods of time, long compared with the famous T, the reciprocal of Hubble's constant. The flashes of light are emitted from the remote galaxy at regular intervals according to proper time there and all the while it recedes faster and faster. As observed here, the flashes are received at progressively longer intervals. Each successive flash has more distance to cover than the one before it and so takes longer to arrive. In Eddington's model, from any receding galaxy, there is a last flash which can arrive here. As we approach that final stage, the intervals between flashes become very long indeed. In order not to become involved with the duration of the flashes, the times during which they are visible, let us reckon the time intervals from the beginning of each successive flash. How long must we wait for the last flash? We must wait during all eternity. And, you may ask, how does that fit in with a finite universe? Well, Eddington has not closed off the time axis, at least towards the future. It is possible to talk about eternity but not spatial infinity. It may also occur to you that what we are now discussing is something analogous to what de Sitter meant by saying that time appears to go more slowly at great distances. With the last flash on its way to us, so far as we are concerned, time has come to a halt over there. Unless we count the flash which will arrive after infinite time, the last flash but one is all we shall ever know of events in that remote region. We might as well be satisfied with the last flash but one, because even if we could contrive to wait eternally long for it, the last flash would have no energy left in it on arrival and there would be nothing to detect in any case. Perhaps it would be better to say that the last flash but one is really the last for us.

Apart from the logical difficulties of trying to cope with infinity, there is a moral in all this. As Eddington once justly remarked; our picture of the world is a consequence of the human mode of thought. It appears that temporal and spatial views do not give the same kind of answer to questions about where the horizon is. The spatial answer can come out in several ways, depending on how we define distance. Does the expansion mean that there are coming to be more miles between us and that distant galaxy or that miles out there are getting longer? Eddington was at some pains to say what he meant by distance (4) and related our proper distance in a specific way to R as an ubiquitous standard. (5) Eddington's idea of distance enabled him to place the horizon a finite way off. In other models, you may find the horizon at differing distances according to the way in which distance is defined and how the course of the expansion runs.

Now for some other conventional features of Eddington's model. The cosmological principle applies because in the static stage it was completely uniform; later all observers could see similar distributions of galaxies and in the end each observer has just one galaxy to see; in a sense, the universe looks alike to all of them. It is also homogeneous and isotropic

---

4. EDDINGTON, SIR A. S., Loc Cit. P. 93 ff.
5. There is certainly nothing absolute about the distance between the marks on that platinum bar at Paris; among other things it depends on temperature. Now that the standard has been redefined in terms of the wavelength of light of a particular spectral line, it is still not absolute.

in the usual cosmological meaning, though here again, towards the end, the sense of the terms becomes rather artificial. At the end each of the isolated island universes runs down to thermo-dynamic death. Strictly speaking, thermo-dynamic death does not mean the situation where there is no energy left in a system but when it is equally shared by all bodies. In this state of equipartition nothing can happen any more as no body has a potential with respect to another. Ultimately the energy will be radiated away into space as "low grade" heat and become lost to the system; there is no longer a source of energy but there is a "sink"; space into which it can disappear. Just as measurable time had a beginning after something began to happen, so it has an end when nothing happens any more.

The cosmic repulsion governs the course of history in Eddington's model and does so in a precise way because of the deep significance which he gave to λ. But for an observer who begins his work after the horizon has become empty there is effectively no λ for expansion can no longer be observed. But from the point of a cosmic super-observer who can see the whole universe, it is still at work and continues regardless of thermodynamic death. Likewise, once the horizon has become empty, R has lost its local meaning too, for it is not the kind of quantity one can learn about on our galactic scale. Space would appear, but for the local presence of matter, as flat as Euclid could have wished. Only a cosmic de Sitter could still evaluate R. The history of things in Eddington's model lets us look at the de Sitter universe in a new way; it is not empty because there never had been any matter in it but because the matter has dispersed to the limit. De Sitter's space is where matter has been, apparently leaving only a geometry of space behind it.

As a representation of the real world, Eddington's model was unsuccessful: it was too deeply committed to specific numerical values, many of them arrived at theoretically by deduction. A specially serious difficulty follows from the revised time scale, involving a much lower velocity of recession. This brings about dimensional changes in the model which necessitate a value of average density much higher than that now derived from observation. (^{7}) It seems unlikely that anyone will try to develop the model further to overcome its difficulties, at least in the near future. With all its problems Eddington's universe remains one of deep intellectual appeal: it has a conceptual beauty and philosophical depth which very few other models can equal.

6. EDDINGTON, SIR A. S., Loc. Cit. P. 68.
7. WHITROW, G. J., Loc. Cit. footnote P. 145.

Abbé Lemaître, also in the 1920's, constructed a relativistic expanding model but with quite a different starting point. Whereas Eddington had preferred a quiet beginning, in his own words "not too unaesthetically abrupt," Lemaître chose a spectacular first event. This question of beginning is always one of great difficulty; scientifically the problem is insuperable, involving a causeless event, although there are ingenious logical ways of evading the question. Eddington separated the beginning of time, by his operational definition of it, from the beginning of matter. Like the ancient Chaos, the static primordial material had been in existence since before the beginning of time and so did not have to be created. This is one way of putting the ultimate question out of reach.

Lemaître supposed that all matter came into existence as a dense *combination* of fundamental particles, a gigantic atomic nucleus. (8) As protons have a radius expressed in our terms as about 10^{-12} cm, and one can assign similar dimensions to other particles, Lemaître could cram all the 10^{54}g of primordial material into a very small volume. At the instant of creation Lemaître's universe had a "radius" R of about 1 A.U., 93 million miles, very small indeed on the present scale of things but not just a point. This primaeval nucleus or "world-egg", as it has been called, was very unstable and immediately broke up, the pieces again breaking up into smaller and smaller portions forming innumerable atoms of matter as we now know it. The process released vast amounts of energy, being the biggest nuclear explosion that could ever happen. So we have particles flying apart in all directions with a wide range of velocities, some quite high, and space increases in size to accommodate them. Now a kind of sorting out process ensues; after some time the faster particles have gone further and in any small region selected out of the whole volume of space the velocity does not differ much from particle to particle. So within a small volume we can say that the matter it contains is relatively at rest. Accordingly, in such a volume gravitational forces can come into effect to form gas clouds out of the atoms in it. Although its expansion has permitted the matter to cool down a great deal by thinning out the intense radiation, it is still hot enough for the particles to have considerable thermal energy, so, as in any other gas, the particles exert a pressure by colliding and bouncing off each other and the cloud does not immediately collapse. Over a period of time these gas clouds can contract locally to

---

8. LEMAITRE. G., *The Primaeval Atom*, reprinted in Munitz Loc. Cit. P. 339 ff. Students may find this paper rather heavy reading as Lemaître's style tends to be obscure.

form galaxies of stars. Here we have an easy explanation for the tendency of galaxies to form in clusters, a cluster represents an original gas cloud which has fragmented. In this case a whole cluster of galaxies can be the unit which participates in the expansion because its members remain gravitationally bound to each other as a relic of their common origin.

After this pause in which the present structure of the universe takes shape, the cosmic repulsion really begins to work and the original outward motions of the gas clouds, which were essentially of constant velocities, are gradually converted into the ever increasing velocities of cosmic recession.

Once the cosmological recession is established, the history of events in Lemaître's model is essentially the same as in Eddington's. Structurally, the difference between the two is in their origins; the one violent and the other nearly imperceptible. In either case the beginning is pushed back a long way; in the Eddington model it is completely inaccessible but Lemaître thought that in the high energy cosmic rays could be found relics of his initial explosion. Today there are quite different explanations for the origin of cosmic rays, which are no longer thought of as electromagnetic radiation but as nuclear particles travelling at velocities close to c and consequently having very great energy.

In other respects Lemaître's model as described is out of date as the average density of matter is now taken to be much lower, even including the mass-density of radiation, consequently R, T and λ would need modification. With adjustments, it could be brought up to date on these points. Unfortunately there would still remain a very serious difficulty. The bursting of the "world-egg" can account for the formation of the heavy elements but these are a very minor portion of all matter, which is observed to be preponderantly hydrogen. It does not seem possible to account for this in the description of the primaeval atom. The real problem here is that the protons, electrons and neutrons were supposed to be *combined* into one gigantic atom, not existing as individual particles close together. When such an atom disintegrates most of it ought to be found as bigger fragments than hydrogen atoms which only contain one proton and one electron each. When complex atoms like uranium disintegrate they do not break up into hydrogen but into much heavier fragments which are mostly quite stable and do not break down further.

 The present leading exponent of "big bang" models of the universe is George Gamow. Like Lemaître he starts with the explosion of a superdense state of matter which he calls the Ylem. In order to avoid questions about how the Ylem came into existence as a superdense mass, Gamow supposes that it was the result of a contraction to the limit from some former less

NGC128, a galaxy of peculiar structure which was formerly thought to represent a collision between two galaxies. It is well known to radio astronomers as the strong radio source Centaurus A. Mount Stromlo photograph, from the 74 inch telescope with an exposure of 90 minutes in yellow light.

NGC55 a spiral galaxy seen edge on, photographed at Mount Stromlo with the 74 inch telescope in 1 hour in red light.

dense state. Nothing can be said now about what it was like before the contraction, nor what caused the contraction; no evidence can survive the Ylem stage which consisted of a featureless mass of protons, neutrons, electrons and radiation. The degree of compression was such that all the matter now to be found in a volume of a few thousand million light years radius would then have occupied a volume a few times larger than the present volume of the sun.⌉

In contraction the Ylem became very hot and one second, as we reckon time, after it reached maximum density the temperature was 15,000 million degrees. Having reached the limit, the Ylem could not stay like that and began to expand. At the beginning the density of radiation exceeded that of material particles so there was plenty of driving energy to start the expansion very rapidly indeed. As the universe expanded it also cooled down at such a rate that after one year the temperature had fallen to about three million degrees.

Gamow has no trouble about getting hydrogen out of the Ylem; it was not a superatom, just a miscellaneous gas of particles. Indeed, his problem was the opposite of Lemaître's; how to get some heavy elements.

Considering the first hour or so of the expansion of Ylem, when most of the atom building was done, there is no problem about hydrogen of atomic weights 1, 2 and 3, nor of helium of weights 3 and 4. But there exists no nucleus of weight 5 nor one of 8, which the balances of nuclear forces do not permit. However, it proved possible to work out thermonuclear processes by which the problem of the missing nuclei can be side stepped, ([10]) and Gamow has a plausible mechanism for the production of the whole known range of elements, supplemented by their production in stars.

Given a supply of matter in gaseous form, Gamow proceeds to explain in considerable detail the processes by which first vast gas clouds and then embryo galaxies and stars are formed. Stars and galaxies in this model go through various stages of evolution, as indeed they must, and the concept of evolution applied to the universe as a whole, becomes the dominant feature of Gamow's cosmology.

Through the remarkable achievements of astrophysicists, a great deal is known about the evolution of stars, virtually from birth as globules of gas to the final slow fading out at the white dwarf stage. According to

---

9. This mass of the universe can be derived from the then current figure for the average density, 10^{-27} g/cm³ and the value of R.
10. GAMOW, G., *The Evolutionary Universe*, Scientific American, Sept. 1956, P. 136 ff. and FOWLER, W. A., ibid P. 82 ff.

Schatzman, who is an authority on white dwarf stars, it may take 10^{11} years for a white dwarf star to cool down and finish its career as a cold superdense lump of matter, like frozen Ylem, which can stay like that for all eternity. As yet we know little about the evolution of galaxies compared to what we have learned about stars, [11] though we can see them at various stages thanks to the finite velocity of light. In Gamow's model, as in most others, the nearest galaxies are seen at a later stage of evolution than those at great distances, which are seen younger by thousands of millions of years. Apart maybe from the enigmatic "quasars" recently discovered and so far of uncertain status, there does not seem to be much difference between galaxies near and far at least as far as the distances to which we can penetrate. This is not altogether astonishing; after all a few thousand million years is not likely to be the greater part of the lifetime of a galaxy. So far investigations have mostly been concerned with their early history and, in most models, we do not expect to see galaxies in extreme old age. However, as Eddington supposed, they may all run down to thermodynamic death in the end. Presumably in Gamow's universe matter which started as a uniform mass of superheated Ylem can finish as small scattered masses of cold Ylem. Something like this would happen in the evolutionary model, because there is no mechanism by which the original contraction could be repeated.

Gamow does not need a cosmical repulsion; the expansion was started by the explosively violent disruption of the Ylem in which matter acquired a motion of recession which simply persists on the large scale. As in the Lemaître model, matter sorts itself out according to its velocity so that, in cosmically small volumes, differences in velocity can be neglected and gravity will account for the formation of the clusters of galaxies. Accordingly, there is a velocity/distance relation in Gamow's model which does not call for a continuing force of cosmic repulsion. In the expanding models considered before this, the cosmic repulsion makes the velocity/distance relation also a velocity/time relation, that is to say, long continued observation would show that the velocity of recession of a body increases as we watch it and, after a long enough time, it will vanish over the horizon.

This point brings us to the question; what is space like in Gamow's universe? It turns out to be hyperbolic, which raises two considerations, of which the first is that it is open. Gamow argues that as the universe

---

11. Oort, J. H., *The Evolution of Galaxies*, Scientific American, Sept. 1956, P. 100 ff.

must be able to expand without limit it must also be limitless in extension. ([12]) This does not mean that the universe is merely unbounded in
his view, it is infinite in extension. Thus it must be actually, not potentially,
infinite. The geometrical consideration arises in this way: like so many
others, Gamow starts on a foundation of general relativity. It will be
recalled that Einstein settled for closed space because of his mathematical
difficulties, but open space was not inconsistent with relativity once other
solutions to the equations could be found in terms of universal expansion.
It is actually possible to find a solution for a model expanding in Euclidean
space, ([13]) but it has some strange features. As Milne showed, the most
remarkable feature is that the model has a horizon which recedes with the
velocity of light, revealing more and more galaxies, a sort of continual
creation of new galaxies appearing to have the velocity of light from which
they immediately begin to slow down.

It is clear that all infinite space cannot be occupied by a finite mass of
matter so there must be an infinity of matter in the universe of Gamow.
He is careful to point out that it does not involve a logical contradiction
to say that this matter can contract to high density and expand to low
density. ([14]) Gamow illustrated his argument by quoting Hilbert's intriguing
proposition about the hotel with an infinite number of bedrooms, all
occupied but in which as many more guests as you please can be accom-
modated. In essence this proposition amounts to what Newton said to
Bentley about infinites being neither equal nor unequal.

As Gamow also says, the question of what curvature space has is one
to be settled in the long run by observation. His model is homogeneous,
isotropic and fully in accordance with the cosmological principle; in a
definable sense, the distribution of the galaxies is uniform. In hyperbolic
space the volume of a sphere increases at a rate greater than the cube of
the radius and, at great range, the number of galaxies ought to increase
faster than the cube of the proper distance. For spherical space the
reverse holds. In principle, counting the number of galaxies visible per
unit solid angle at various large distances should answer the question.
The trouble is that we have to look as far as we can see, maybe even
further and these are just the distances at which the images on the
photographic plates are too faint to give definite measures of luminosity
distance. Moreover, we still do not know what corrections ought to be

12. GAMOW, G., *The Creation of the Universe*, Mentor MD 214, 1957, P. 42 ff.
13. WITROW, G. J., Loc. Cit. P. 108 f.
14. GAMOW, G. *Creation* Loc. Cit., P. 35 f.

applied to allow for the changes of luminosity in galaxies in the course
of their evolution. The answer is not yet clear.

Before the time scale was revised, T, the so called age of the universe,
was a very serious problem for Gamow. By much ingenious juggling with
the observations he could get a velocity/distance relation of 180⫽km/sec/
mpc, giving a T of 3.4 x 10^9 years with which he could make do. Now
the new time scale of more than 10^{10} years suits him very well.

In the pragmatic sense this evolutionary model of the universe is rather
successful. It does give a plausible account of the observable universe and
it is sufficiently flexible to accommodate itself to substantial changes of
observed values. It also provides a fairly detailed account of the formation
of stars and galaxies. On the whole it is at least as successful as any
theory yet constructed. On this basis, it is reasonable to ask why anyone
should look for another theory. A good reason is that it may be possible
to construct other models which are equally successful but more pleasing
in respect of inaccessible events. Gamow chose an initial contraction from
which he could get a rebound into expansion because it was mathematically
simple. The criterion of choice here is really an aesthetic one and does not
have the compelling force of a fact of observation. You may, if you are
so inclined on religious or other grounds, start your model with the
creation, by an external agency, of matter in the Ylem form. In either case
the Ylem stage is what mathematicians call a singularity, which may be
described as an abrupt change in an otherwise smooth curve. If the smooth
curve is a representation of physical events there is a tendency to dislike
singularities because in large scale phenomena abrupt changes are either
inadmissible or have very drastic effects, at least that is the way it goes
in terms of man-made laws. Accordingly some cosmologists, of whom
Eddington was one, prefer to avoid "bangs". Others, like Gamow, do not
object to a singularity but do not care for the implication, which Lemaître
accepted, that the singularity represents the creation of all matter and
energy out of nothing. This very difficult topic will come up for discussion
later.

The rebound of Gamow's Ylem was so violent that gravity could not
at any stage overcome the expansion which must go on for ever. ([15]) But
it is possible to choose other conditions in which the expansion is limited.
Mathematically these conditions are expressed in the way $R(t)$ runs with
time for the various combinations of k and λ. There are a number of

---

15. This has the consequence that the old Newtonian problem of gravitational
potential in a uniform infinite system does not arise.

possibilities: one corresponds to Einstein's static model, others expand indefinitely at various rates while still others alternately expand and contract. These several possibilities were tabulated some thirty-five years ago by Lemaître and have been classified and displayed in a comprehensive set of diagrams by Bondi. ([16]) It will be noticed that the curves of $R(t)$ (Fig. 9) for the oscillating models start from close to zero, rise to a finite maximum and fall again nearly to zero. It appears that oscillating models may be finite in respect of space and material content, but the duration of each cycle is in some sense finite, [though there does not seem to be a way

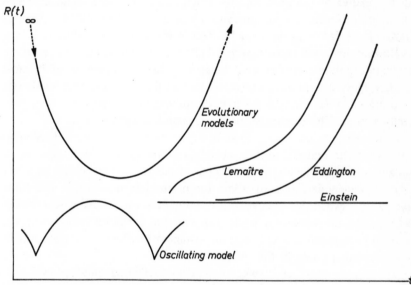

Fig. 9   Some examples of how R changes with time in various models. Curves not to the same scale.

to measure it.] Although to the satisfaction of their constructors, these oscillating models seem to push the creation back into the infinite past, it looks more like a conjuring trick than anything else, for the singularity at the beginning of a cycle is surely a discontinuity in time and reference to "previous" cycles must be meaningless.

In a conventional oscillating model, like that of Opik, ([17]) the universe

16. BONDI, H. *Cosmology* Loc. Cit P. 80 ff. They can also be found in a more summary form in NORTH, J. D., Loc. Cit. P. 130 f. To represent Gamow's model more accurately the curve would have to be drawn more like a V than a U to show the singularity.
17. OPIK, E. J., *The Oscillating Universe*, Mentor MD 289, 1960. Chapter 29.

+ There may be now.

is of necessity presented in the phase of expansion and at a fairly early stage because [there is no clear evidence of slowing down.] Such evidence would take the form of observations that the velocity/distance relation nearby differs from that at a great distance where we see an earlier stage. An oscillating model thus presents the appearance of a finite relativistic model with a prediction that instead of expanding asymptotically into a de Sitter model, it will, sooner or later, start to contract.

Oscillating models fall into two groups which can be described as "deep" and "shallow." In the deep models the contraction phases bring all the matter of the universe into a stage like Gamow's Ylem in which all structure formed in the previous phase is destroyed and the universe starts off its career again virtually from a fresh creation. Given a means of slowing down the expansion to a halt so that contraction can set in, an oscillating universe of this kind looks very like a succession of Gamow models. Obviously nothing can be said about the structures and conditions in previous cycles beyond the hypothesis that they must have been generally similar. The singularity of $R(t)$ precludes anything else.

In the shallow oscillating models $R(t)$ runs in a smooth wave-shaped curve and the material system need not come to an Ylem stage. By a suitable choice of factors it is possible to make a model which does not contract far enough to bring about the destruction of material structures like galaxies when $R(t)$ comes to its minimum value. In this case it would be possible for observers fairly early in an expanding stage to "see through" the minimum and at great distances to observe galaxies still in the contracting phase of the system.

An obvious question is: what makes an oscillating universe oscillate? In the physical sense this can come about only through the inter-play of opposing forces. Clearly they cannot have radically different laws like gravitation falling off inversely as the square of the distance and the old cosmic repulsion increasing as the distance, for that would permit no reversal. A possible combination is gravitation and pressure. In a Ylem stage where radiation preponderates over matter at the enormous temperatures involved, the pressure of radiation can be the driving force to overcome gravity and thus set off the expansion, but later it falls away with distance faster than gravity which can then slow down and ultimately reserve the expansion. In circumstances like this there seems to be no reason why the general motion of a material system should not be reversible.

In the ordinary sense of reversible processes, like stretching a rubber band and letting it spring back, there is no difficulty about conceiving

† Note p 153.

expansion and contraction. Many of the equations describing physical phenomena are symmetrical with respect to time; if you put in $-t$ instead of $+t$, the equation is still perfectly good and it describes the opposite sequence of events, rather like running a cinema film backwards. But, symmetry of equations or no, there are many processes in our world which we can be reasonably sure are irreversible. As Eddington said and Popper more recently; [18] "the arrow of time flies only one way." In a contracting universe the arrow of time flies the same way as it did while the universe was expanding. In space-time talk; the world-lines which used to diverge are now converging but there is no suggestion of going back along the t-axis.

Radiation provides an example of processes which are irreversible in our world. A broadcasting station sends out radio waves into a large volume of space, some of the energy, but only a tiny fraction, is caught by radio receivers and most of the rest is absorbed by the earth and its atmosphere, finally to be degraded into heat. Try to imagine the reversal of this process even on the basis of the radio receivers alone. Every receiver would have to know the right instant to send back its received energy to the transmitter in proper phase and sequence to reunite there as it was before transmission. Just like running the film backwards, it is one thing to talk about advanced potentials but quite another to get nature to play this kind of game, at least on the large scale. [19]

At this stage it may look as if we are headed for a discussion of entropy, but that is not the intention. Entropy is a concept developed here on the local scene, in the presence of the rest of the universe and here its expression in the second law of thermodynamics [20] works extremely well. Eddington and his contemporaries took the second law very seriously and insisted that in accordance with it their universes must run down. But there are now models to which the law does not apply. This seems to be true of all infinite models which are not the closed systems in terms of which

---

18. POPPER, K. R., *Nature*, 1956 March 17, Vol. 177, P. 177, P. 538. This is a letter to the Editor and is followed by correspondence between Popper and others to be found in Nature, Aug. 18th, 1956, Vol. 178 P. 381 f., Nature, June 22 1957 Vol. 179 P. 1296 f. and Nature, Feb. 8, 1958 Vol., 181, P. 402 f.
19. Microphysics is quite another affair: sub-atomic particles may not be interested in the direction of time. Thus on occasion a positron may be an electron going backwards in time. See WHITROW, G. J., *Natural Philosophy of Time*, Nelson, 1961, P. 280 ff.
20. From one point of view it amounts to saying that you cannot expect heat to transfer from a body at low temperature to one at a higher temperature. Consequently you cannot make a "perpetual motion of the second kind" which involves precisely that juggle with heat. The law being statistical does not say that this is absolutely impossible, just that it is utterly unlikely.

entropy is defined. It is a large extrapolation to assume that it applies to a universe as a whole and no one has undertaken to prove that it does. Let us be content to say that if someone wishes to make a model of a universe repeatedly contracting after expansion, he must make sure that he is not involved in irreversible processes. In many cosmological theories it is very convenient to assume that the second law does apply.

An endless succession of expansions and contractions implies that the system does not lose any energy. At first sight it seems obvious that, by definition, the universe at large cannot lose energy though it may be degraded; but in terms of oscillating models we must look a little more closely into the implications. Does it necessarily follow that a contraction of the material system will bring about a contraction of the system of radiation? In other words one has to guard against the situation where we could have a contracting "universe of matter" and a still expanding "universe of light." If that happens, each cycle of the material universe will lose a substantial amount of energy in the form of radiation and over a number of cycles the oscillation will die away. To have a perpetual oscillating universe it is essential to reverse the dispersal of radiation as well as of matter. Then we can be sure that the energy of the system, upgraded again by the Doppler effect in the contracting phase is fully restored.

There are many other models based on general relativity which we have not attempted to cover, some of them very remarkable indeed, like the finite rotating model which Goedel built upon his solutions of the equations of relativity. But enough ground has been covered to show some of the problems and the achievements of relativistic cosmology. Relativity provides the best methods so far developed for handling problems on the medium and large scales needed for cosmology; certainly no better approach to the problems of gravitation has yet been developed, despite its complexity and difficulty. This constitutes one of the two reasons for seeking alternative theories and the other is concerned with philosophical problems lying at the deepest roots of relativity, which some acute thinkers find have been treated unsatisfactorily or not at all. So far it has not been possible to unite relativity and quantum theory, though we can hope that some day this will be done, with the two theories meeting on the middle ground. Then Eddington's vision will have become reality. Meanwhile we must take another path leading to models based on different ideas. For a while there will be more emphasis on time than on space.

The Bird of Time has but a little way
To fly and lo! The Bird is on the Wing.

CHAPTER 13

ANOTHER APPROACH

The relativistic cosmologies sketched in the last preceding chapter can be called theories of space. Time and matter figure prominently enough but it seems as if space, personified as it were in $R(t)$, dominates the scene. One did not wish this impression to prejudice the outlook on the models; hence the remark that de Sitter's space is where matter has been, as if the Cheshire cat of "Alice in Wonderland" had vanished, leaving his grin behind. There is a stage of appreciation when a space-grin without there ever having been a cat can be rather disturbing.

There are many concepts of space ranging from a mere set of relations, which seems to be inadequate for cosmology, to a sort of ether which is more than what is strictly necessary. The influence of the past remains strong, tempting us to see, in the spaces of relativistic cosmology, subtle kinds of ether. Despite protestations to the contrary, there are probably many scientists, including some knowledgeable about relativity, who keep some such notion at the back of their minds. It is not hard to see how the concepts of space-time and fields form some sort of equivalent to the old ether. One can even see that a strong motive behind the development of field theory is the natural objection to action at a distance. Such vaguely ether-like notions work well enough on the large and medium scales; it seems to be only when one has to deal with the microcosm that space-time and fields become excessively difficult to use. Perhaps the truth is that these ideas are not relevant at all to things on the smallest scale. So far as cosmology is concerned there is no absolute right or wrong about the use of these ideas, it is rather a question of being consistent about concepts. In relativity the seeming physical reality of space stems from the way in which the rules were laid down and should be recognised as the appearance that it is. ([1]) It is perfectly legitimate to treat matter not as the cause of kinks in space but as the manifestation to us of the kinks, but there is no point in doing it unless some contribution to knowledge can be gained.

---

1. NORTH, J. D., Loc. Cit., discusses the concepts of space and length in Chapter 16.

It is only natural that the exploration of a field of thought like relativity should reveal still other tracks to follow and equally natural that questions should arise about the basic structure of the field. As Bondi has expressed it:

"However, it is clear that there is room for wide divergence of opinion. This in itself is no disadvantage, since serious progress is impossible in a state of self-satisfied unanimity." (²)

A leading questioner of the bases of relativity appeard in the 1930's in the person of E.A. Milne. Coming to cosmology from mathematics, Milne had a preference for deductive methods and a strong sense of conceptual propriety. Combining these with a powerful capacity for original thought he soon became a centre of controversy. (³) Milne came to the conclusion that time is fundamental rather than space, so we can describe his cosmology, based on the kinematic relativity which he developed, as a theory of time.

The deductive method, widely used in mathematics can be simply described. We start from a minimum number of general principles and develop detailed propositions by exploring their consequences. This works admirably, for instance in developing the geometry of Euclid from a few simple postulates but it is quite another matter to make the method work in terms of the universe. To do this properly involves an intellectual re-birth, banishing from the mind or at least positively identifying and setting aside everything one knows of the empirically determined laws of science. The objective is to construct laws from first principles. Obviously great vigilance is needed to avoid unconsciously shaping the process of deduction so as to arrive at results agreeing with empirical laws already known. The majority of scientists are opposed to the deductive method, being convinced that it is impossible either to obtain a guaranteed set of first principles or to carry out the process of deduction in practice. Indeed, there have been some spectacular failures, like Descartes, in attempting the deductive approach. Most cosmologists today believe that Milne also failed, but it is generally agreed that he made some very important contributions to the science of cosmology. For the purposes of this course it does not matter whether or not Milne succeeded in making a workable representation of the universe. Let it be said quite loudly; there is not and

---

2. BONDI, H. Loc. Cit. P. 5.
3. It was not easy to be neutral about Milne. I recall an argument across a dinner table which was rather suddenly ended by a leading astronomer, Professor B. Lindblad, with the remark "Milne was a genius." That was nearly 30 years after the concept had been published and Lindblad was not a disciple.

probably never will be a wholly successful model of the universe. What we are interested in is the kinds of ideas people have about the universe and how they come to form them. If you do not let empiricism or some other philosophical attitude make you resentful, a study of Milne's work can be both interesting and rewarding.

Right at the beginning of Milne's considerations are questions of epistemology and he overtly chose operationalism: to put the idea crudely, what you cannot at least in principle, do or observe, is without meaning and questions about such things are unanswerable; therefore not legitimate. If you wish to define something then you must do so in terms of an operation which is at least conceptually possible. This attitude clears, or rather removes, a lot of ground. Thus Milne would not consider a true de Sitter universe as legitimate for it cannot contain any observers and so nothing meaningful can be said about it. There is, for Milne, just one universe and we are in it as observers; whatever happens, happens in the presence of the universe at large. This is in flat opposition to the outlook of many thinkers. Bertrand Russell insists that we consider as many universes as there are possible distributions of matter. (4) Let us adopt for present purposes Milne's operationalist outlook and argue about its merits some other day.

Having swept so much away it is necessary to establish some firm ground on which to begin building. It is not permissible to use any of the existing quantitative laws of science or ideas like distance and space; these must all be derived from a minimum set of principles and observations treated in terms of them. We shall now engage in an exercise to show that this can be done. It will be necessary to appeal to observations which are conceptually possible rather than practical and, though this may seem highly artificial, please bear in mind that we are dealing with ideas, not things and it is adherence to the rules that counts.

In a way reminiscent of Descartes, let us begin at home with ourselves as conscious beings who can observe by means of light. However little else we know about the universe, it obviously has light. For the sake of simplicity let us at first consider only one observer and allow me to provide the observations, with Milne as my guide. Whatever I observe is light which affects my eyes. (5) These incidences of light are registered

4. BONDI, H. Loc. Cit. P. 3 ff.
5. I do not propose to enter here upon a discussion of sense-data and objective reality. It has been taken for granted that I am in a real universe and am setting out to learn about it.

in my memory as a sequence and I can identify them by means of the series of integer numbers, starting with any particular light incidence that I select. This sequence appears to be unalterable; there is no way to rearrange it. The light incidences, which I call events, are related to the numbers in such a way that I say event 14 happened after event 13 and event 15 happens still later. As the order is unalterable, it follows that what I call time goes only one way. It is always getting later, but of course there is no saying how much later. This counting of events is rather tedious so let me have a clock. This is just a device in which there is a succession of events which, in principle, I could count and compare with the light-events I have been numbering. The hands and dial on the clock simply save the effort of counting. The use of a clock involves making two observations together, seeing a light-event and seeing the hands of the clock. Please do not ask about the rate of the clock, this is not a meaningful question.

As there is light in the universe, perhaps you will allow me a lamp with a pushbutton switch so that I can make flashes of light. The possibilities of observation now increase very remarkably; frequently some time after I have made a flash I again perceive light. Repeated flashes assure me that there is some connection between them and some of my subsequent perceptions of light. It seems that there are sources of light which are independent of my lamp and others which are not self-luminous but only become luminous after the lamp has illuminated them. Further observations show that it takes longer to evoke a response from some than from others. In other words, if the lamp flashes at a clock reading t_1, there is an observed response at t_2, but the number of clock-events between them may differ very considerably. In fact, it is possible to arrange some of these light responses in an order corresponding to their $(t_2 - t_1)$.

It is possible to describe the difference between these light responses by saying that some are later than others, but it is easy to get this confused with that unalterable increasing lateness called time. Provisionally let us call it the t-factor. Next, by using a series of flashes in rapid succession it appears that some of the responders have changing t-factors. Observations with series of flashes on responders with constant t-factors show that, on the average, I get my series of responses later by the t-factor but with the same interval from flash to flash as they were originally made. The conclusion is that these responding things do not wait a while after being illuminated to respond, they do so immediately. It seems that it takes time to illuminate something and it is reasonable to suppose that the reverse process, the responder illuminating me, will also take some similar time. On this basis it is possible to refine the statement of the t-factor. What

is characteristic of these responders I take to be $\dfrac{t_2 - t_1}{2}$ and for simplicity

of notation let it be called d, so:

$$d = \frac{t_2 - t_1}{2}$$

If you wish, you may equate my d with one of the concepts you call distance but I am not interested in that, I am content to call it the delay. All I can tell you about d is that *it is what light does in time.*

I can also locate in time events with d factors. The time identification of such an event is:

$$T = \frac{t_2 + t_1}{2}$$

with reference to the origin of time arbitrarily chosen at t_0. Following Milne this can be called the epoch of the event in my scale of time. At this stage the event in question is the illumination of a responder and I can place it meaningfully in the time sequence of observed events, even though it is a present observation of a past event.

This d is clearly different in some way from time which changes only one way: d is variable in two senses, for it can be different for different responders and for particular responders d can change with time. Moreover changing d can become larger or smaller. At this stage the technical vocabulary can advantageously be enlarged. If a responder has variable d I say it moves relative to me. ([6]) With my crude clock I can assign some rough rate of change of d. This I call speed. Some responders have speeds higher or lower than others. Constant d is zero speed. As d is what light does in time, I can in some sense say that light has a speed.

Now, during the course of these observations I have often observed several responders simultaneously. In other words, they were together(in time)but in another sense they were not. This not-togetherness, which I call separation, is also variable in two senses, rather like d. It can differ between pairs of responders and I have an impression that in terms

---

6. In the days of Greek Science the term motion was used in this way, meaning any kind of change. The words now being introduced are not to be taken to have any wider implications than directly indicated. Thus speed strictly means rate of change of something with time, the word is thus often used with no relation to distance.

of particular pairs of responders, it can change in time. If you let me make a device resembling what you call a protractor, it will now be very useful. I shall make it in this way: at any instant there appears to be a greatest separation which my eyes can perceive and this, marked on my protractor, I divide into arbitrary steps. The protractor saves a lot of concentrated effort in observing and remembering. Indeed, it is a sort of memory, a record of the steps of separations that I estimated before. As a memory it enables me to find the separation of responders which are not simultaneous, using a continuous light source as a reference.

I suppose, as a pure hypothesis, that the self luminous objects also have a d characteristic, but there does not seem to be an easy way to check this so far. Progress can be made in another way. It does look as if the property called separation is something like the property called d. As remarked earlier, it has a similar freedom of variability and starting with responders of known d, it is possible to calculate hypothetical d's which represent separation. (7) It seems that d may not be exclusively a relation between responders and me, it may also be a relation between responders. By a refined experiment with a pair of responders with the smallest d factors I can locate and with a moderate separation, I find that by illuminating one of them, I can observe a subsequent and much fainter illumination of the other. Knowing d for both of them, I can extract from the total delay corresponding to the equivalent d of separation, and it works out as if there really is a d between them. In thus coming to the conclusion that separation is of the same nature as d, the use of a protractor was not fundamental, it was only convenient.

Having already acquired a primitive notion of direction from the separation of events with, in the first instance, the same d factor and observable simultaneously, I can now elaborate the idea and so arrive at the notion of d in several directions. Without going into all the detail you can see how, along these lines I arrive at a concept of distance and of space, how I can assign three dimensions to space and only one to time. In short, I can talk meaningfully about *where* as well as *when*, all based on one primitive concept of time and some observations.

The next step is to set you up as observers at some of these responders, armed with clocks, lamps and those glorified protractors called theodolites. The telescope part of the theodolite will come in handy for observing the readings of each other's clocks; the lamps will also serve a second

---

7. You may call this a sort of trigonometry if you wish.

purpose as signalling devices for exchanging information. At first there will be no general agreement about the results of observation, but as between pairs of observers with constant separation it is easy to make comparisons. Repeated observations of each other's clocks will enable pairs of observers to find the relation between their time scales and it will then be possible for one party to "regraduate" his clock so that it agrees with his friend's clock in the sense that they can get the same results for mutual observations. Since any regraduation of a clock will alter the distance scale dependent on it, the two observers can now agree about measures of distance between them. It will be a little more complicated for the case of two observers in uniform relative motion, but still quite possible for them to come to an agreement about observations of each other. In a still more complex way something could be done for two observers in accelerated relative motion.

The questions to be answered about the chances of agreement between more than two observers at a time will come up later. At this stage it is better to pause and take stock of the situation and to examine again the objection that was thrown at Milne and his colleagues. In all that has been done here the observers and observations are alike hypothetical and the air of reality which the description lends is purely illusory. While this may indeed be true it is irrelevant. There never has been a suggestion that what we have been talking about should be taken literally as practical operations. It is not reasonably possible with a lamp and a clock to measure some distance in terms of the transit time of light. In fact when Milne proposed this conceptual scheme in 1935, there was no prospect of any experiment resembling the notion. In 1945 it was common practice to measure hundreds of miles by radar; by 1955 radar had reached to the moon and by 1965 to the sun. A conceptual experiment had, in 30 years, become practical for distances in the order of 100 million miles. By this argument it is not suggested that we can ever work radar methods for very distant bodies, but it is demonstrated that Milne's approach was conceptually valid. The radar set does not differ in principle from a lamp and clock. The differences between them are of degree not of kind. Unless you have found some new logical rule whereby you can nominate a stage when a difference of degree becomes a difference in kind, you must not deny the validity of this deductive chain of reasoning. It would be a good idea now to read Milne's own account of these matters. (8)

---

8. MILNE, E. A., *The Fundamental Principles of Natural Philosophy*, Munitz, Loc. Cit. P. 368 ff.

You will notice that we have departed in respect of details from his treatment but, it is to be hoped, not from the rules.

None the less you may still have some reservations about this whole chain of argument. Indeed it has been developed at some length for the very purpose of showing where the dangers lie in trying to derive established ideas *de novo* from first principles. For instance, looking back a couple of pages to the responders characterised as moving; in the absence of concepts of distance and space, was it strictly justifiable to assert that repeated responses came from one and the same responder? There are a number of debatable points of this kind which add up to a serious doubt as to whether it is ever really possible to put oneself intellectually in the position of an observer totally ignorant of all that has been done in science and yet be capable of carrying out such a process of deduction. Milne's deductive approach produced an astonishing array of results, from complete theories of gravitation and dynamics to new theories of the atom and of electromagnetic phenomena. Enough will have been seen already to show that the starting point of it all was cosmological. ([9]) It had to be, because the very act of discarding all the quantitative empirical laws leaves nothing else. Naturally, in broad terms, the laws of science so deduced give much the same results as the conventional laws, but the different method by which they were formulated provides new possibilities for exploration. Many new relations and previously hidden connections are now revealed. It was by no means all plain sailing to carry out the task that Milne undertook and many of the problems were of a new and unfamiliar kind.

In earlier chapters some distinction was made between the laws of motion and actual motions; it had no more value than pedagogical convenience but now it must be abandoned, since the laws of motion to be deduced can be no more than descriptions of actual motions. In the absence of an organised body of laws to be extrapolated to the largest scale, it is necessary to appeal to the cosmological principle to introduce the necessary element of law and order, to allow us to treat the universe as capable of rational description. In other cosmologies we have seen the cosmological principle as a broad general indicator of how things must be, but we did not attempt to apply it in detail for we know quite well that observers in differing local circumstances would get vastly differing views of the universe. The view from our earth must be very different indeed from that of an observer in the vicinity of the Orion nebula, and

---

9. Ibid, P. 354 ff.

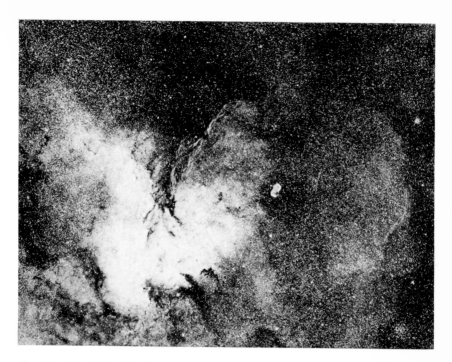

Gas Clouds of the Southern Hemisphere. The Southern Milky Way is rich in emission nebulae, some of which are shown on this photograph made with the Upsala Schmidt Telescope located at Mount Stromlo Observatory.

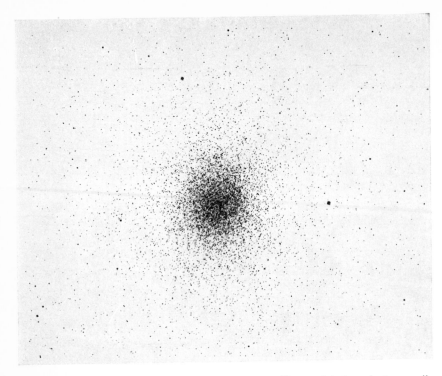

NGC5139, better known as ω Centauri, a magnificent globular cluster, easily visible to the naked eye. It is one of the globular clusters, forming the halo of our galaxy and is something like 20,000 light years distant from us. In this photograph from Mount Stromlo Observatory, the stars appear dark against the background as is the case with nearly all plates used by astronomers; the reversed bright-against-dark is essentially used for purposes of illustration.

from that of another peering through the dust clouds of the Coal Sack. Even though his lines of sight may be clear, an observer whose motion differs much from the general radial motion of expansion in his part of the universe will get an odd perspective of things. Thus if he has a high velocity he will see the light of bodies ahead of him bluer and of bodies behind him redder than they would be seen by observers attached to galaxies who are thereby partaking in the motion of expansion.

For kinematic relativity the cosmological principle must apply, not in a vague general sense but strictly and in detail. From what has already appeared in relativity according to Newton and according to Einstein, it is important to be careful about defining the classes of observers to whom a principle of relativity applies. Let there be no mistake about it: the cosmological principle here is a principle of relativity in a broad and powerful sense. Relativity says that the laws of science are the same for the specified observers; Milne says that the observations are the same and it is precisely these observations which will settle the laws of science in due course of deduction. As a matter of fact whether certain people, like Dingle, are dubious about it or not, the cosmological principle is in the nature of a principle of relativity and is entirely compatible with any such principle which has so far been formulated. In the strict sense in which Milne had to employ it the cosmological principle finally states that all equivalent observers will arrive at identical descriptions of the universe. This is more rigid than his earlier statement quoted by North that:

"Not only the laws of nature, but also the events occurring in nature, ... must appear the same to all observers, wherever they be, provided their space frames and time scales are similarly oriented with respect to the events which are the subject of observation."

The class of observers to whom the cosmological principle applies in detail are called fundamental observers and, of course, they must be equivalent. Equivalence has to be expressed in a different way from that used in earlier chapters, where it turned out that equivalence meant having the same laws of science, which guaranteed getting the same results from the same observations. In this kinematic approach the only possibility of equivalence we have yet seen is between two observers with clocks. When we have succeeded in getting two clocks at a distance to agree about a series of observations by regraduation, they are said to be congruent. In extrapolative science they would be said to go at the same rate. One would be inclined to think that one observer out of a pair could make himself and a third observer equivalent to one another and thus make the second equivalent to the third. Unfortunately this proposition applies to equalities but not to equivalences. Basically, equality implies the possibility of superposition, that is to say, having the alleged equals together *here* and *now* to establish their equality; but equivalence is meaningful in terms of *there* and *then*. Milne found out that it was possible to have a linear equivalence, as it is called, between three or more observers only when they have relative motions of a special kind. As is well known, every observer finds himself at the centre of a spherically symmetrical observable region and each observer's centre is as good as that of any other observer. It seems to follow that the only kind of relative motion which will satisfy the conditions and so permit of general equivalence is a uniform radial motion of expansion, so we are left with the proposition that fundamental observers are those taking part in the motion of expansion.

For quite a long time Milne's theory was the only one in which expansion is inevitable; general relativity permits either expansion or contraction and the choice, when building a model, has to be determined by observation. If we consider a swarm of moving particles located at some instant within a certain spherical volume and observe them from its centre, after some lapse of time they will be found to be dispersing. Imagine, if you like, some gas of such low density that collisions between the molecules are very unlikely and they are accordingly too widely separated to influence one another. Molecules originally moving away from the observer will continue to do so; those orginially approaching will pass by the observer and so, sooner or later, will be moving away from him. The swarm of molecules will occupy an ever increasing volume and, as time goes on, will sort themselves out in distance according to their velocities, so a simple velocity/distance relation appears. This dispersal of a swarm of particles can be used rather effectively as an example of an irreversible process. From a strictly cosmological point of view, we now see a way to dispense with cosmic repulsion, not by virtue of an explosion, an appeal to brute force, but from a simple and lucid principle. If, after this, we still want to retain λ in our equations it will be out of deference to the mathematical aesthetics of generality, not out of necessity.

The fundamental observers are conceptual creatures for whom we way provide equally conceptual abodes, called fundamental particles, mounted on which they trace out in space-time the non-intersecting world-lines of Weyl's Postulate, diverging from their common origin in the past at $t = 0$. The system of fundamental observers, fundamental particles or their world-lines or the frame of reference which the system constitutes, is called the substratum. As there can be a fundamental particle at every point of the substratum we can consider it as a continuous background for the universe. The uniform relative motion of the expansion is the only motion of the substratum and results in a velocity proper to every point in it, differing infinitesimally from the velocities of neighbouring points. Such a motion is said to be hydrodynamic because it can be dealt with by the same mathematical formulae as the motion of a uniform fluid in which there is no turbulence. It is thus to the substratum that the cosmological principle applies in its strict form; here the equivalence is complete. At any time t the observable part of the universe for any observer comprises a finite spherical region of the substratum, of radius ct, which he sees from the centre.

There is another kind of particle to be found in Milne's universe, the class of free particles. These do not partake of the motion of the substratum but may have non-uniform motions and thus accelerations relative to fundamental particles. However these free particles are not quite so free as their name might lead one to think. Milne showed that there have to be equations of motion for free particles because their motion is always executed in the presence of the rest of the universe and so must be describable in the same or at least in convertible terms by any plurality of observers. Without going into details, you will be interested to hear that free particles have a strong tendency to cluster round fundamental particles and come to rest relative to them with the result that observers on them are likely to find that they are to all intents and purposes attached to fundamental particles. Free particles do not seem to remain free indefinitely.

There is a natural tendency to equate the nuclei of the galaxies to fundamental particles as Milne himself did. Some rather strong objections have been made to this identification, obvious and desirable as it seems. As appears from the consideration that there should be a fundamental particle at every point, they must constitute a continuous set; the nuclei of the galaxies, even if infinite in number, cannot be more than a discrete set. ([10]) The question thus arises why only some fundamental particles should have materialised, which is easy enough to see from the point of view that galactic nuclei are rather big ones to appear at each point. But then even a single proton is too big in just the same sense. Then when one has to consider swarms of free particles collecting around every point it becomes fairly obvious that the fundamental particles of the substratum are an

---

10. BONDI, H., Loc. Cit. P. 137.

idealisation which one must not take literally enough to make into a material structure of stars.

Upon thinking this over a little, you may be inclined to ask if it is really necessary to treat the galactic nuclei as fundamental particles. In a finite volume of space at some particular epoch there will certainly be an infinity of points but only a finite number of free particles which will come to acquire, according to Milne's rules, the velocity of the substratum in their vicinity. These in the fulness of time can coalesce into masses which behave more or less like fundamental particles but are only to be found here and there in a discrete set. This view may be easier to accept for our present purposes but it is well to bear in mind that it too may result in awkwardnesses.

The essence of the whole problem seems to be the necessity to have a strict formulation and application of the cosmological principle at the conceptual level when the laws are being deduced, yet having to account by the same principle for the behaviour of gross matter. In other words we must find a consistent way to loosen the formulation, as Bondi puts it, [11] when describing the world as observed.

The regraduations of the clocks which qualified the fundamental observers may also permit of synchronisation, so that they all read the same epoch. In this case there will be a cosmic time common to all fundamental observers. A simple proportionality between velocity and distance means that when we trace back the diverging world-lines there will ultimately be a time when all the fundamental observers were together in the same place. This is the zero of time, the beginning of the universe, from which the cosmic time runs.

From all this it looks as if Milne produced a model of the universe with perpetual expansion at a constant rate from what amounts to a point origin; and so he did, but it was a model with a dual aspect, either finite and expanding or static and infinite. This will bear investigation before we look at details of the model. Milne was not suggesting that the universe of observation is static, but he does have to deduce a set of laws corresponding to those of Newton, the first law in particular. Though not impossible, it is extremely difficult to do this in the time-based expanding model. Remembering that regraduating clocks changes the scale of distance and that the velocity is proportional to distance, one can see that by using a suitably variable time scale one can transform away the velocity and treat the model as static. It may not seem very pleasing to

---

11. BONDI, H. Ibid.

have an additional time scale that changes continuously as it goes, but it solved the problem in hand and proved useful elsewhere as well. Mathematically the trick is simple enough, Milne just had to make two time scales logarithmically related. Somewhere in the past there has to be a moment, when clocks graduated according to the two scales, agree in rate and epoch. ([12]) The two time scales were identified as t-time and τ-time according to the rule:

$$\tau = t_0 \log (t/t_0) + t.$$

In the t-scale the zero of time $t = 0$ was only a finite time ago at the convergence of the world-lines, but the logarithmic relation puts that moment on the τ-scale infinitely far back in the past; this is the scale on which the fundamental observers appear to be mutually at rest. It is also called dynamic time and it appears that in it dynamic processes used to go much faster than they do now. This was convenient in the days before Baade's revision of T because it completely avoided the time scale problem. The other original scale, t-time, has become known as atomic time as it seems to be the one appropriate to electromagnetic and atomic phenomena. It is of course that in which we made all those light observations to establish the concept of distance.

Coming back to the expanding aspect of the model and treating events in terms of t-time, the motion of the substratum is a uniform relative motion and it is possible to use the Lorentz transformations if space is Euclidean. On the question of curvature Milne said:

"The answer is that he may map his observed events in any type of space he chooses; and having chosen a type of space for measures in one scale of time, he may find it convenient, when he changes his scale of time, to adopt a space conformal to the first as his new map. In the present case, in the simplest description of the expanding system, each observer uses a "flat" Euclidean space of which a portion only is occupied by the system. ([13])

Having made this choice it follows that the space of the static aspect in τ-time has to be hyperbolic or negatively curved. As mentioned earlier, each observer is at the centre of a space which constitutes for him the universe. Previously we have used the term observable universe but here, in deference to our operational standpoint, a stronger phrase is needed. To be still more careful, let us say that each observer is at the centre of a spherical volume of the substratum, of radius ct. As is the nature of the

12. MILNE, E. A., Loc. Cit. P. 361 f.
13. MILNE, E A., Loc. Cit. P. 362. See also BONDI H., Loc. Cit. P. 129 f.

substratum, it continually expands so that the volume which the observer can survey steadily grows and its spherical surface or border recedes at the velocity of light. To use Milne's own words:

"From the point of view of the observer who is central, the external physical space occupied by the enlarging sphere is continually recruited by the gain of space created by the expanding light wave which started at $t = q$. Light creates space, as I said before."

The recession of the border at the speed of light is something that must happen when there is a simple velocity/distance relation and you look to the theoretical limit of vision; nor is it a velocity of recession merely approached asymptotically, the distance ct is quite finite. For the observer, this horizon is quite impenetrable and really limits his universe but we know that this is the same for every other observer and their horizons may overlap in some cases. Thus for each observer, the widening of his horizon reveals always more galaxies. Within his private universe the average density of matter inside a finite volume around him, large compared with the size of a galaxy but small compared with his total universe, decreases with time. On the other hand this density, on the largest scale, increases with distance so that at his horizon it reaches infinity. (^{14}) The horizon is completely impenetrable for him, being occupied at every point. Paradoxical as it may seem, this novel distribution arises from Milne's firm adherence to the principle of the conservation of matter. Without it he would have been faced with a continual coming into existence of new matter, but now it is merely the revelation of matter previously unobservable and so then virtually nonexistent. It must be emphasised that no influence can cross the horizon, causal or otherwise, but what has been happening out there may one day be seen when time enough has passed. Here we have the opposite of Eddington's universe of increasing disconnection, kinematic relativity gives increasing connection.

Upon reflection it will be seen that there is nothing incompatible in these two rules for density, but it is appropriate to ask whether the paradox of Cheseaux and Olbers can arise. From the expanding aspect it will not, because of the red shift brought about by the expansion. In the static aspect the relation between the two time scales does the trick; though the frequency of light does not change with the passage of t-time,* Hence light which has been travelling for a long time becomes reddened and the paradox is avoided.

---

14. Milne's diagram illustrating this notion is reproduced in MUNITZ, Loc. Cit. P. 355.

* it does with τ - time.

It is interesting to note that in Milne's expanding model, because of the precise uniform velocity/distance relation, there is no need for the Doppler factor to appear twice in the formula for luminosity distance as it does in models based on general relativity. There is no difficulty in interpreting the observed redshifts in this sense and in this respect the model does not conflict with observation; indeed it may fit better.

The two time scales do not anywhere cause embarrassing conflicts with observation. Indeed before Baade's revision, it was a great help that the "age of the universe" T comes out in τ-time to provide the greater dynamic age of the universe required. Of course the problem is no longer a serious one and it can be said that the difference of rate between the two since the revision would now be smaller. Possibly the most remarkable observational evidence for the existence of two scales would be to detect the slow change in the universal constant of gravitation which Milne's theory predicts. Put it another way: if we observe for long enough it might be found that all pendulum clocks gradually slow down compared to atomic clocks. In any case it would take millions of years on either scale to produce a noticeable difference.

In our day a more balanced outlook on kinematic relativity prevails than when it was being developed. In science as in other human activities the truth is neither here nor there, and there is something to be said on both sides. It does seem that the strong criticism which Milne provoked led to his adopting an aggressive attitude and consequently overstating his case in some respects. On the other hand some of the criticism was, to say the least of it, ungenerous. The task which he undertook was one of enormous difficulty, indeed it was too much to expect any small group of men to carry it through to success in a few short years. In a course such as ours it is not practical to go deeply into the questions which kinematic relativity raises and once again it is not for us an important question whether or not Milne succeeded in making a workable model of the universe. In a purely pragmatic sense, the kinematic model, despite its many attractive features, involves too many difficulties for anyone at this late stage to try tinkering with it here and there in an attempt to make it into a successful model. The problems are more conceptual than representational, for, if the conceptual structure is sound, the resulting model ought, within the limits of observational facilities, to conform more or less to nature. It has been held that Milne never managed to bring his concept of distance into terms of convertibility with distance as we customarily measure it, yet he did make use of Hubble's velocity/distance relation which was based on conventional concepts of distance. But he did arrive at a velocity/distance relation of great simplicity and conceptual elegance and, moreover, his methods are not so far removed from what astronomers really do as might appear on the surface. Perhaps this problem could be solved by defining the standard metre on a radar basis. Whether agreeing with him or not we certainly are much in debt to Milne for bringing into the open so many hidden questions, assumptions and possibilities. [15]

Perhaps for us his most outstanding technical achievement was the development of the cosmological principle which he raised to a status and endowed with a power never before contemplated. Whether or not you like the deductive method and the operationalist point of view it has to be admitted that they brought some astonishing results. Let Milne have time and light and he will show you a universe full of fascinating ideas.

We have now seen a scheme of things in which the age of the universe is much more than a half fictitious backwards extrapolation, rather a real measure in terms of which "constants of nature" as we have been accustomed to call them, are slowly changing. There have been other cosmologies in which this kind of thing happens though on a very different basis and we may take a brief look at a minor theory framed by P.A.M. Dirac in 1937, in which things change rather fast. Dirac is not a leading cosmologist but is justly famous for his work in microphysics. It was this background which led him, like Eddington, to contemplate those mysterious big numbers and their coincidences, but from quite a different point of view. Where Eddington saw order and necessity, Dirac saw pure coincidence in time. It seems that at one stage Dirac was doubtful whether the universe can really be explained in rational terms [16]

Out of microphysics comes a very small unit of time having the form e^2/mc^3, where e is the electronic charge. This natural unit has the value 9.5×10^{-24}, sec. and its ratio to the then accepted Hubble age of the universe, 1.8×10^9 years works out about 10^{39}. This number to which Dirac assigned the symbol p can be called the age of the universe in natural units. The number p^2, like Eddington's N of magnitude 10^{79}, represents the number of protons and neutrons in the universe while $1/p$ represents the average density of matter. [17] In the same order of magnitude as p

---

15. If the possibility of new kinds of forces suggested by Milne (Ibid P. 367) is sound, one could reasonably expect them to be discovered some day. What can happen usually does.
16. Here I am relying on my memory of something written by Dirac about 30 years ago which I have not since had an opportunity to look up. It is fairly consonant with Bondi's comment (Loc. Cit. P. 160) that Dirac's argument may be called a "counsel of despair."
17. BONDI, H., Loc. Cit. P. 161.

is the well known ratio of electrical to gravitational forces between electron and proton. Dirac suggested that this ratio and $p_?^2$ which represents the mass of the universe in terms of the mass of a proton, are functions of the age of the universe and hence vary with time. If the mass of protons is not to vary with time, then their number must increase as the square of the time since it is represented by p^2. It also follows that the "constant" of gravitation must vary as $1/p$. It seems that Dirac felt there were compelling reasons why the proton mass should be taken as a constant. This argument leads to the conclusion that new matter in the form of protons (and electrons, since matter is electrically neutral on the average) or neutrons must be continually coming into the universe. As he was unwilling to accept this conclusion, Dirac later evaded it by appeal to a variable time scale resembling that which Milne had used and to a possible variation of the proton mass with time. ([18]) In terms of the old value for T this scale makes the age of the universe in natural units $T/3$, about 6×10^8 years, which is remarkably short compared with geological time scales, to name only one. The implication is that natural processes went very much faster in the past, resulting in spectacular changes in what we ordinarily take to be constants of nature. Even on the revised scale where $T/3$ would be 4×10^9 years, the changes are still rather fast.

We shall not follow Dirac any further and will merely mention a strange theory based upon his ideas by Jordan. Where Dirac could not accept the idea of new matter coming into existence, Jordan adopted it and elaborated a model in which new matter appears as quite sizeable lumps in the form of stars or even whole galaxies. ([19]) The only point of real interest for us in theories of this kind is that out of the investigation of time comes a suggestion that matter may originate in time as well as before time. When Milne found a proposition of this kind appearing out of his own mathematical manipulations he was quick to point out its purely formal origin. None the less, however much it may seem like a self evident truth, it is an assumption to hold that all the matter now in the universe came into existence at the same moment of the remote past or that it has always existed in some form for past eternity. In theories of matter we can explore the consequences of different assumptions.

18. NORTH, J. D., Loc. Cit. P. 201 ff.
19. BONDI, H., Loc. Cit. P. 163 f and NORTH J. D., Loc. Cit. P. 205 f.

What! Out of senseless nothing to provoke
A conscious Something . . .

CHAPTER 14

THE THEORY OF MATTER

When we say that matter has physical existence several ideas are implied and some others, though they can be shown to be relevant, simply do not occur to us. Not wishing to venture into the land of the thousand meanings, one may be content with a fairly unsophisticated view of what existence signifies. We do mean that matter exists in space and in time, *somewhere and somewhen*. In terms of space-time it has world-lines and if these world-lines lie outside our light cones we can only find out about matter there by means more indirect than any kind of observation. But even within our light cones, by what observations could we ever learn about matter if it did not have both spatial size and duration in time? These quantities cannot be zero: perhaps they cannot be smaller than certain minimum units, like Dirac's atomic unit of time. (^{1})

We derive our idea of the continued existence of matter from observations of it in gross form. A lump of matter can be observed from time to time and recognised, so, with good practical reason, we suppose that it continues to exist while it is not being observed. The idea of continuity is very strong here but it does not appear incompatible with atomicity of time in terms of a minimum significant unit of duration. Long ago an idea was developed that time proceeds by discrete steps and in the twelfth century Moses Maimonides developed it to a stage where he could assign a duration to these instants. In modern notation it would be about 10^{-9} sec. Apparent continuity of existence involved a recurrent act of God, re-creating the whole universe for each instant. The logical basis of this concept is to permit change to take place. Otherwise change seemed logically impossible because it would mean change by no steps whatever.

At this stage let us confine our attention to questions of duration. So far as we know, there is no upper limit to the duration of matter. We are

---

1. Apart from its shape e^2/mc^3 this unit can be seen as the time (Chronos) for light to cover the smallest significant distance (Hodon) which is around 3×10^{-13} cm., the range of nuclear forces. With $c = 3 \times 10^{10}$ cm/sec., the result is 10^{-23} sec., close enough to Dirac's 9.5×10^{-24} sec.

acquainted with matter about 4 x 10^9 years old on this earth and, however much its form may have changed and still be changing, it does not seem to have any innate tendency to go out of existence. ([2]) It may be objected that radioactivity and nuclear fission, natural and man-made, must result in some loss of matter in the form of radiation; though it must be admitted that this is true, it is also insignificant on the cosmic scale. Fortunately for us, stable atoms enormously outweigh the unstable.

Light, by which is to be understood electromagnetic radiation in general, has physical existence also, but the manner of its existence differs interestingly from that of matter. Although for certain purposes light can be treated as consisting of particles, there is no suggestion that photons are material particles in the same way that atoms or their constituents are material. To say that light endures in time is meaningful in the sense that we can observe now light which we are sure left its source at an earlier moment. Without this there would have been no foundation for relativity.

Light can be observed only once. Any observation we can make of light results in its ceasing to exist. Light entering the eye is absorbed in the material structure there and the energy which it contained is used to give us the perception of light; what we perceive is not, but has been, light. This is equally true of the detection of electromagnetic radiation of any frequency, as in the reception of radio signals or observing gamma rays by their effects on matter. It seems that light can exist only for so long as it is free to move. This is part of the meaning of what was said in the last chapter; distance is what light does in time. ([3]) The existence of light is terminated by its absorption in material structures, but apart from this, there seems to be no way for light to go out of existence. Unless the accepted interpretation of the red shift is sadly astray, we know that light is arriving here which has already endured for something like 5 x 10^9 years. In specifying this period, in talking about the endurance of light in any way we are using a concept of time as we know it but this does not mean that photons have the same kind of time. If we could have an observer with a clock moving at the velocity of light, that is on the world-line of a photon, he would find there, where $ds = 0$, that his proper time is always zero and so, from his point of view, what we take to be a light journey of 5 x 10^9 years, takes no time at all. This is a very strange idea for us, but it follows inevitably from special relativity: Photons exist on the

---

2. This has to be taken in a statistical sense. Protons and neutrons do not have any individual identity in nuclei.
3. It is also another way of saying that $ds = 0$ for a light ray.

border of that half-world outside our light-cones. They are the fastest things we can see in time.

The origin of light does not appear to be a complete mystery; it is born of matter. There are several processes for the begetting of light, summarised in turning on a lamp, which only superficially affects the luminous atoms as the energy which they radiate is supplied to them from an external reservoir. Some other processes yield light from deep within the atom, from the nucleus itself, in the form of gamma rays. (4) There is a series of processes involving matter by which light in its whole range of frequencies can be created.

But what about the inverse process; can matter be made out of light? Einstein's equation shows how much light of a given frequency would be needed to make so much matter, but it does not say anything about practicability beyond letting us see that on conventional scales, an enormous amount of energy has very little mass.

The creation of matter out of radiation seems to be a very difficult process. A photon of sufficient energy can create a pair of particles, an electron and its antimatter counterpart, a positron, but only by interaction with already existing matter. There does not appear to be a mechanism by which photons can "condense" into particles of matter in empty space. (5) Except by appeal to some so far unknown process we must say that matter cannot be created purely out of radiation.

Of these two kinds of physical existence that we have learned about through observation, we can account for the existence of light in terms of matter but not for matter in terms of light. It is natural to ask, how then did matter come to exist? But is does not necessarily follow that questions about that are answerable, nor that when an answer can be given it will be very satisfying. Let us review some of the answers as to when matter came into existence.

ANAXIMANDER — It is always coming into existence, new vortices are always forming and old ones are unforming. The Apeiron has always existed.

---

4. There does not seem to be a way to convert the whole of a nucleus into radiation except, in principle, by bringing it up against its counterpart in anti-matter, when both will be annihilated. This possibility can be demonstrated with electrons and positrons. Perhaps fortunately, there does not seem to be much anti-matter in our neighbourhood.

5. Beware of Newtonian type light corpuscles here. If photons were like that you could imagine somebody like Clerk Maxwell's Daemon cramming about 10^9 green photons into a volume of 10^{-13} cm radius in 10^{-23} sec in an abortive attempt to cook up a neutron. Professor Heisenberg, to name only one of the spokesmen for science, has declared that kind of magic "black."

LUCRETIUS	— It is always coming into existence by accretion and wearing down again. The atoms have always existed.
NEWTON	— It came into existence in the distant past.
KANT	— It started to come into existence in the distant past and is still coming in. But in any particular region the process will terminate in the future.
EDDINGTON	— The question is unanswerable. The primordial material has always existed, it is only possible to say when it began to expand.
LEMAITRE	— About 1.8×10^9 years ago.
GAMOW	— Before the Ylem, thus before 3.4×10^9 years ago, but it is impossible to say how long before.
MILNE	— At the zero of time when the world-lines began to diverge. If a numerical value is wanted it is necessary to choose a scale.
BONDI	— It is always coming into existence.

There is not very much agreement among these views except trivially, that at least some matter has existed for some time in the past. As a matter of fact the majority vote among most groups today would be that all matter came into existence in the past, but among them would be plenty of scope for argument about the first moment, if any, of its existence. [6] Taking an opposite view we have Anaximander, Lucretius, Kant and Bondi, with propositions that matter will come into existence in the future. Anaximander, Lucretius and Kant made this fairly easy for themselves by providing a primordial something out of which to make matter. Only Anaximander and Lucretius and Kant will allow the unmaking of matter, to return whence it came.

All this seems to leave Bondi, and some others who think along the same lines as he does in a different position from the post-relativity cosmologists we have discussed so far. There is quite a group of these people, Bondi, Hoyle, Gold and McCrea, to name the most prominent, who have worked along several lines with theories of matter, which are, for technical reasons, called steady-state theories. The ultimate basis of these new theories is the cosmological principle. You have already seen how Milne could use it as a large scale principle of relativity, which indeed it is, though disguised as a description of the material universe. The cosmological principle provides that if observations are conducted in

---

6. NORTH, J. D., Loc. Cit., P. 389 ff.

different parts of the universe at the same cosmic time, or if there is no cosmic time, at local times relatable in some way, ([7]) then the results will be the same.

But even from one observing position the aspect of the universe will change in time in expanding models because the average density is always diminishing. If you make a cosmic clock out of average density, in so doing you have already implicitly appealed to the cosmological principle. Working in terms of spatial sameness always leads back to the ultimate singularity, the coming into existence of matter. Now there is a way to avoid this quite simply by having the cosmological principle work temporally as well as spatially. We have seen that a purely spatial principle brings in difficult temporal problems, so let them be abolished; let the universe present the same aspect from all places at all times. This is the Perfect Cosmological Principle, which lies at the heart of steady-state theories.

If, on the large scale, the aspect of the universe is unchanging in time, it is in a steady state. This does not mean static, for it is not the undifferentiated sameness that for Eddington amounted to non-existence. Let us distinguish three major large scale situations. There is the static state in which nothing ever happens. Then there is the evolutionary state of relativistic cosmology in which things happen for a while. Finally there is the steady state in which things are always happening. ([8])

Bondi, Hoyle and the others are well aware that the universe expands and that it cannot remain in a steady state if the average density of matter decreases in time. The only way in which this problem can be solved is by the continual appearance of matter in the universe at such a rate that the density remains the same despite expansion. ([9]) The name of this process, continual creation, has caused rather a lot of unnecessary misgivings and argument largely because the word creation has some theological overtones. Earlier in this chapter the phrase, coming into existence, has

---

7 Like knowing the stages of evolution of galaxies and thus being able to recognise comparable times. And with anything like a point origin is that not a perfectly good way to set a cosmic time?

8. The kinematic relativity state is intermediate; things have only been happening for a while but they will never stop happening in the future.

9. Your attention should be drawn to a charge made by Dingle (Scientific American, Sept. 1956. P. 236. (The italics are mine.) "The cosmological principle, however, *alters observed facts* to make them accord with its requirements. Observation appears to show that the density of matter in the universe is continually decreasing. This cannot be so, says the principle, unobservable matter must be in the process of creation out of nothing in just the right amount to keep the density constant". This is very dubious logic and not up to Dingle's usual standard.

been used but will now be dropped and the word creation will be used, without overtones.

Another objection which has often been raised is that continual creation involves abandonment of the established laws of conservation of matter and energy. Perhaps in the customary formulation of the law of conservation of matter something has been left unsaid. As usually stated it is really only appropriate for the creation in the infinite past. For those who take the view that creation was completed in the finite past, it should be put in the form: Matter can *no longer* be created or destroyed. Looked at in this way the conservation law does not appear as strong as it has generally been considered to be, for there must have been, in many cosmologies, a time when it did not apply.

As you have seen, there is nothing very novel in the idea of continual creation and, where there is a primordial material, it is probably a more useful idea than completed creation in the past. Anaximander and Lucretius balanced it with continual uncreation, which, in a model without expansion could lead to constant density, and provides also a pleasing symmetry. It is exceedingly difficult in the context of modern science to start with a primordial material out of which matter can be fabricated. Radiation does not look at all promising for the purpose because of the lack of a mechanism for converting free energy into the more or less permanently fixed form of matter.

There are some alternatives: one suitable for completed past creation is to push it back to the infinite past. Then it can be said that creation did not happen because however far back you look, matter has always existed. ([10]) Then the problem of equipartition of energy arises: the universe would inevitably have run down to a standstill in the infinite past.

The constructors of oscillating models seek to avoid the thermodynamic problem by providing for periodic contraction to start the sequences of events all over again. However, it is not clear that this will work during infinite time. All periodic processes known to us, apart from those supposed to take place in the interior of atoms, die away in time due to dissipative losses of energy.

Another approach which is applicable to continual creation has been used by Hoyle and by McCrea. Hoyle inserted a "creation field" into the field equations of general relativity. Modern field theory has come a long way from the classical luminiferous ether. While a field now looks at first sight like a pure abstraction, the theory works astonishingly well, on

---

10. NORTH, J. D., Loc. Cit. P. 395 ff.

the basis that what is mathematically possible may also be physically possible and what is physically possible usually happens. That is not the way field theorists describe what they do, but that is how it often seems to an outsider. ([11]) For us the problem is not how to get matter out of a field, but the origin of the field, which is as much a problem as whence comes primordial material.

McCrea's zero point stress as a source of matter is a far more abstract sort of an idea than Hoyle's C-field and we shall not attempt to discuss it. However, he has introduced a most interesting notion that particles of matter can produce new particles by a kind of breeding process. ([12]) Not only does this provide a field mechanism for continual creation but more, it points to a way in which a very deep problem might be answered. The question is why are all the basic particles of matter of each sort indistinguishable from one another? There is evidence that matter, electrons, protons, hydrogen atoms or any other kind of atoms are exactly the same here and everywhere and at all times. It is not just a case of making them from the same blue print with the usual small production tolerances: of each kind the particles are completely the same, totally indistinguishable one from another. It looks as if there is the equivalent of a perfect cosmological principle for the microcosm and it applies in detail. On the other hand it is obviously possible to distinguish between lumps of matter containing considerable numbers of atoms. The various kinds of atoms themselves are clearly distinguishable as to chemical properties though they are made up of particles completely without individual identity; it is the pattern that counts. From the human point of view there is nothing to choose betweeen one atom and another of a particular isotope of some element, but when many atoms are brought together to constitute a lump of matter, it can be specifically identified because the arrangement of atoms can be sufficiently permanent for that purpose. Thus it is relatively easy to distinguish one diamond from another, even though they are very similar. Identity can only be established when there is a possibility of labelling.

There is a curious question which arises no matter what origin of matter one chooses to adopt; that is the problem of anti-matter in which the signs of the electrical charges are reversed. Anti-protons have negative charge and anti-electrons, (positrons), have positive charge. In all other

11. HOYLE, F., *Recent Developments in Cosmology*, Nature, Vol. 208, 1965, No. 5006, P. 113.
12. McCREA, W. H., *Why Are All Electrons Alike?*, Nature, Vol. 202, 1964, No. 4932, PP. 537 f.

respects anti-matter is exactly like matter and, from a distance, there is no way to tell which is which: for all we know the Andromeda galaxy might be made of anti-matter. It certainly exists, for physicists have had some thirty years experience in dealing with positrons and more recently anti-protons have been produced in nuclear experiments. The quantum theory does not discriminate in favour of one of the two kinds of matter; the only awkward thing about having a quantity of anti-matter here is that it and an equal quantity of matter would immediately annihilate each other in a violent explosion with the release of all their content of energy. Whether one considers creation in the past or continual creation, the same difficulty arises; for the creation of anti-matter ought to be equally as probable as the creation of matter. As a matter of science it is not very satisfying to make an appeal to pure chance for the polarity of all the matter in the universe but, in any theory in which all matter was once in a dense state, as in the models of Lemaître and Gamow or even Eddington, there cannot have been anything like similar proportions of matter and anti-matter. So far the alternative seems to be a theological explanation.

Continual creation does not offer a solution to that problem, but at least new matter comes into existence in a diffuse state, rather than concentrated. Perhaps McCrea's "breeding" process could be adapted to lead to regions of matter and anti-matter more or less evenly distributed about the universe.

Now we come to a difficult question, the creation of matter out of nothing. This one has aroused more antagonism than all of the others put together, it has been called an appeal to magic. ([13]) If by magic we are to understand anything observable for which there is no known scientific explanation, then in the long run the existence of matter is magical in all theories. Putting in primordial materials, fields or what you fancy, only postpones the question by one step. Do you remember the ancient scheme where the earth was supported on four elephants which in turn were supported on a tortoise? If, like Bondi, we adopt creation *ex nihilo* we will be no worse off and will have saved some rearguard fighting. It also leaves you free, if you so desire, to bring in some overtones on the word "creation."

The phrase, creation out of nothing, contains more implications than are immediately obvious, though you may be inclined to think that some of the points verge upon hair-splitting, this is indeed a subject in which choice of phrase can be rather important. Objection has been taken to

13. NORTH, J. D., Loc. Cit. P. 403 f.

saying of the creation of matter, that *it comes from nowhere* and an alternative has been used by Prof. Hoyle: *it does not come from anywhere*. Upon reflection it will be seen that the former of these two is logically objectionable; it uses what logicians call a spatial predicate, implying there is a nowhere, a sort of *limbo* outside the universe, whence comes new matter. Let there be no doubt about that issue; Professor Bondi is not talking about some process of conversion, not about creation out of chaos nor out of fields nor out of energy but simply out of nothing. It is a process for which science can offer no explanation whatever. For an explanation, for a cause, you would have to seek elsewhere than in science.

Those who object to magic in the sense that we have just used the term, seem to think that science must inevitably be able to explain everything. It amounts to saying that everything has a cause and the causal relation can always be rationally stated. This seems to be a rather outmoded view of science and one might better cut off the chain of cause and effect with a causeless event or a series of events of creation *ex nihilo*. Otherwise how does one avoid an infinite regress in the search for a first cause? ([14])

Bondi takes the view that continual creation enables more questions to be answered than does creation completed in the past. You may wish to know whether these are questions which might be asked about any scheme of creation or whether they are peculiar to continual creation; if exclusively the latter, perhaps not as much would be gained as might appear. On the whole it seems that these questions could be asked about past creation but no evidence remains on which most of them might be answered.

We certainly have no direct observation of the creation of new matter in our part of the universe. This is a very good reason for discarding Jordan's model in which completely new stars or galaxies of the suddenly appear from nowhere out of nothing; the process would have to be much more inconspicuous than that. It is known that there is far more hydrogen than any other kind of matter so it is reasonable to suppose that the new matter comes in as hydrogen or something that can readily become hydrogen; since big lumps are out of the question, let it come in atom by atom. In order to keep the average density constant against the expansion, it would have to appear at a rather specific average rate determined by the observed density and red-shift. The rate works out at about 10^{-46}g/cm^3/sec., which is one hydrogen atom in a litre of space in 5×10^{11}

---

14. There is endless scope for debate on these points but it seems more suited to armchairs and flagons round the fireplace than to a rather austere lecture room.

years. ([15]) There is obviously no possibility of observing any such process on the local scale.

The creation rate is specified as an average rate, but it should not vary much from place to place. If existing matter had an influence on the creation rate in its vicinity, say causing it to be greater, we would expect the masses of large accumulations of matter to grow rather quickly, but there is no evidence whatever that this is the case. It is simplest to assume that existing matter has no influence, leaving the rate essentially the same everywhere.

Returning to the form in which matter is created, there appear to be three possibilities in Bondi's theory. On grounds of simplicity hydrogen atoms are preferred though if protons and electrons were to appear separately but in substantially the same numbers in any region they would sooner or later combine into hydrogen. Equality of numbers is necessary to prevent the appearance of net electrical charges; matter is, on the whole, electrically neutral. The third possibility is the creation of neutrons. Unlike neutrons within the nuclei of atoms, if indeed they can be said to exist there as neutrons, free neutrons are unstable and, with a half-time of some 20 minutes, they decay into hydrogen atoms, releasing in the process as radiation the surplus energy which makes them unstable. Although the neutron is a single particle, Bondi and Hoyle preferred the creation of the hydrogen atom containing two particles because of its lower content of energy: creation should preferably not be a wasteful process. ([16])

The next question is that of velocity and direction. If the new hydrogen atoms come into the universe with high velocities relative to existing matter it would be equivalent to its appearance at high temperatures. ([17]) This does not seem to be the case as the average temperature of matter outside of stars and other massive bodies appears rather low. Thus it is reasonable to suppose that new matter is created with velocities not differing much from those of matter already in the regions where it comes into existence, which are rather low. In the terms introduced by Milne, it partakes of the motion of the substratum. Newly created matter looks more like fundamental particles than free particles. Similar considerations apply to direction: if new matter were to be created with motion in some

---

15. BONDI, H., Loc. Cit. P. 143.
16. It is interesting to observe the tendency to apply rational considerations even to events without assignable causes.
17. When we speak of the temperature of gas we are in fact talking about the velocities of its particles. In the case of a solid body, to say that it becomes hotter is to say that its atoms are vibrating more vigorously. In a gas they have more freedom and do not have to remain in a fixed arrangement.

preferred spatial direction, we should expect to find an overall streaming of matter in that direction. In this case the best supposition is that the initial direction of motion of new matter is random. There do seem to have been some answerable questions raised by continual creation.

We have described continual creation as going on in time and as taking place everywhere in space and this raises a question about how it stands in relation to space-time. The theory of relativity as distinguished from cosmology knows no preferred directions in space-time, but continual creation quite definitely does imply a preferred direction, because the quantity of matter is always becoming greater as time passes. Bondi points out ([18]) that this incompatibility of continual creation and general relativity need not be disturbing: all cosmology involving non-static models has this same incompatibility as the cosmological arrow of time flies only one way. This indeed is what Weyl's postulate means; to make a model of the universe using relativity, a direction of time in space-time must be chosen.

We have now arrived at the stage where we should survey some features of a steady-state model of the universe. The two principle versions of the steady state are due, one to Hoyle and the other to Bondi and Gold. Professor Hoyle used general relativity for his starting point and modified the equations to include the creation field. Bondi and Gold, who saw from the work of Milne, that there are other bases for a modern cosmology, have developed their model from the perfect cosmological principle in a deductive way. Of course the two models are much the same in general aspect, but it will be more instructive for us to follow the line of thought taken by Bondi.

The steady state universe with its unchanging aspect cannot, as a whole, run down to equipartition of energy, by virtue of continual creation and so will remain on the large scale as we see it for all eternity; its future is infinite. But what of the past? The perfect cosmological principle demands that it has always been the same and this means that it has already had an infinite past. Thus we cannot assign a date for the beginning of continual creation, which thus never had a beginning. As you know there has been a strong dislike in many quarters for an elapsed infinite time. Now that we are really faced with an infinite past, let us again examine the objection to it. It seems to be based on the notion that future eternity can never be consummated; it is like counting in natural integer numbers, it is always possible to think of a later date. Choosing *now* as a starting point, any future date we can nominate specifies a finite future

---

18. BONDI, H., Loc. Cit. P 150.

time beyond which it is still possible to go. If one argues that it is just as possible to count backwards with integer numbers; -1, -2, -3 and so on, as to count forwards, there is a suggestion that this will not work for time because past dates are actual. The basis of this suggestion seems to be human consciousness. We have a specific knowledge of events that we have experienced whereas we can only guess about future events. The future is indeterminate whereas the past is fully determined. One would be more interested in this suggestion if it could be shown that specific knowledge could, even in principle, be had of all past events. Actuality does not really seem to constitute a valid objection to past eternity. There need not be a first date for the steady state universe.

As continual creation has been going on for infinite past time, the universe must contain an infinite quantity of matter. It is a sound practical rule of local science to suspect that, if our calculations show some physical quantity like mass becoming infinite, something has gone wrong. The cosmological situation is not quite so clear-cut. It is not the *existence* of an infinite quantity of matter that counts but its *local* influence. Thus it is clear that in the steady model no observer will be confronted with the paradox of Cheseaux and Olbers. The expansion takes care of that. Later we shall see that there is a kind of horizon for each observer from beyond which neither casual influences nor information can ever reach him.

To avoid any misunderstandings about the meaning of distance it should be emphasised that in this theory the concept of proper distance is used throughout. This means that the measurement mesh has a constant scale and does not grow with distance. Instead of the miles out there getting bigger, there come to be more miles between us and distant bodies as the expansion goes on. Thus when we speak of a constant average density of matter in the observable universe, we mean that the amount of matter in a certain proper volume remains constant. In conventional relativistic models the density diminishes with time in proper volume but remains constant in co-ordinate volume which grows with the expansion. Accordingly there is a basis for saying that in the steady state theory matter, or to be more precise, mass is conserved since the mass of matter per unit proper volume is unchanging in time. In this theory the number of galaxies per unit proper volume is quite definite. ([19])

---

19. Ibid, P. 144. Note that while the mass is conserved in a proper volume it is clear that if we could label the individual particles, continual creation and expansion would result in their ultimate movement out of that volume and replacement by others. Bondi sets out the formula for the number of galaxies per unit volume on P. 147.

In the steady state the amount of radiation in any large region must also remain constant. Stars formed through the condensation of new matter are the sources which keep up the level of radiation on the large scale which would otherwise decrease because of the loss of radiant energy consequent upon the expansion.

It is easy to see from observation that our part of the universe is very far from a state of thermodynamic equilibrium. It appears that there is something like 100,000 times as much matter as there is radiation, the one estimated at 10^{-29} g/cm^3 and the other at 10^{-34} g/cm^3. Both of these are rather rough figures. In the ordinary course of things we can only observe distant matter by its emission of radiation. In practice it is technically rather difficult to detect cold matter by its absorption of radiation and there might be more of it lying around than we think, but if there were enough of it to change the proportion of matter to radiation by a factor of 10 we could expect to have gravitational evidence of its presence. Equally it seems unlikely that the flux of radiation could differ by as much as a factor of 10.

Even if the overall ratio were to change by a factor of 100 in consequence of new discoveries there would still remain a gross unbalance. Here, there and everywhere matter is radiating away energy to be engulfed by space which has an apparently limitless appetite for it. There is no suggestion whatever of any radiation being rejected or returned by space nor is there any evidence of equipartition of energy amongst matter on the large scale. In evolutionary theories with creation completed in the past the sources of energy must ultimately run down and in some of them all the radiant energy will have been swallowed up in the depths of space. The present large disparity between matter and energy is then a function of the present age of the universe; you will recall how in the Gamow theory there was at the very beginning of the expansion a severe reverse disparity. The density of energy was greater than the density of matter, enormous as that was.

In the steady state theory the present disparity remains the same at all times. It does so because continual creation is always bringing in new matter and thus forever renews the sources of energy. And what happens to all this energy? It goes into the expansion of the universe and in performing its part is run down by the Doppler effect so that it cannot accumulate.

Our large scale observations of matter are essentially concerned with the galaxies. Perhaps we should say that the particles of the substratum are newly created hydrogen atoms but, in practice, we must treat the galaxies as the observable members. For any fundamental observer the number of galaxies in some finite part of the universe which he can observe will remain substantially constant no matter how long he observes. During an extended set of observations, lasting, say, 10^{10} years, he will have noticed that distant galaxies formerly in that region have moved out of it but they have been replaced by new ones forming from continually created hydrogen. [20] In any large volume galaxies fully formed occupy only an insignificant fraction of the space and there is plenty of room for new ones to form between them.

It follows that in as large a region as can be observed there should be galaxies of varying ages, from those so new as to be just recognisable, to rather old ones. Statistically there must be an average age which works out at $1/3\ T$, which, on the recent value of the red-shift, is 4×10^9 years. [21] However, there must exist in the universe galaxies of all ages and the age of any particular galaxy can be found only by observation. Our own, even from terrestrial information, must be at least 4×10^9 years old and may be considerably older, perhaps as Bondi suggests about 10^{10} years.

There is a problem in all cosmologies, mentioned before, that we do not yet know much about the evolution of galaxies. They are seen to exist in a wide variety of forms and have quite a range of sizes. On the local scene we find small galaxies of no particular shape or discernible structure, such as the Clouds of Magellan which are in fact satellites gravitationally bound to our own Milky Way. Other irregular galaxies of different sizes are to be seen at large distances. Spiral galaxies such as ours and Andromeda, some tightly wound, others rather open, are to be seen at all distances. All the spirals have rather dense bright nuclei while there are large classes of what are called spherical and elliptical galaxies which seem to be all nucleus. Two small elliptical ones are attached as satellites to the Andromeda spiral.   .

It has commonly been supposed that there are considerable numbers of very small galaxies both irregular and elliptical and that those which we see attached to the Milky Way and to Andromeda have been gravitationally captured. Upon reflection, there seem to be two things against this. Just as there are limits to how small a mass can form a viable star, there are difficulties in seeing how very small independent galaxies can be

---

20. It may be that in the steady state theory the life-history of a galaxy is longer than in other theories as it gains matter, consequently energy, from hydrogen created in and around it.
21. BONDI, H., Loc. Cit. P. 149.

formed. Then the very small ones nearly all appear to be satellites. A recent theory due to H. C. Arp suggests that these small satellites are in fact fragments expelled from the nuclei of big galaxies during explosions which occur in them. Thus the Clouds of Magellan may have been born out of the nucleus of our galaxy something like a hundred million years ago.

The spiral galaxies which are near enough to be examined in reasonable detail all have a halo of globular clusters surrounding them and presumably so do the more distant ones. These star clusters are not supposed to be products of Arp's explosion theory but are an integral part of spiral galaxy structure. They are so arranged as to make the whole a spherical system with the spiral arms as an equatorial belt. Some of the globular clusters belonging to our own galaxy, of which there are about a hundred known, are visible to the naked eye, notably the magnificent ω Centauri which must contain something like a hundred thousand stars.

There is a variety of forms intermediate between spirals and ellipticals. While it is quite straightforward to classify galaxies according to shape, it is quite another matter to interpret this in terms of age. ([22]) It is natural to assume that the irregular galaxies are the youngest, but this is by no means certain: they may be examples of the results of disintegration rather than the process of formation. In the steady state theory there must be at places in the universe galaxies which are infinitely old, (unless the fate of aged galaxies is disintegration and dispersal, for which there is no observational evidence.) We have no idea what such a galaxy would look like if indeed it could radiate light by which to perceive it, in fact, within our own practical range of observation, there do not necessarily have to exist galaxies of enormous age; the expansion may have taken them out of range of our instruments.

It is obvious from what has gone before that space in the steady state theory is open, consequently it cannot have positive curvature so that in a metrical formula k would have to be -1 or 0. Starting with the metric for the kinematic model developed by Robertson and Walker, Bondi shows ([23]) that $k=0$; accordingly three dimensional space is flat. Further the significance of $R(t)$, in which R changes continuously with time, is that the distances of galaxies from a fundamental observer increase with time.

---

22. There is a well established theory due to Professor J. H. Oort that spiral arms spread out of and return into galactic nuclei periodically with a time of a few hundred million years, which is cosmologically rather short, being accomplished in a few rotations of galaxy.

As the velocity/distance law must be unchanging in a steady state universe, R is an exponential function of time. Accordingly $R(t)$ can be replaced in the metrical formula by the group exp $(2t/T)$, in which t is a conventionally chosen zero of time and $1/T$ is the value of Hubble's Constant derived from observation. [24] Written in Cartesian coordinates the metric thus becomes:

$$ds^2 = dt^2 - \exp(2t/T)(dx^2 + dy^2 + dz^2)$$

which is the same as the metric for the empty de Sitter universe of relativity. Here there is no need for the formula to signifiy emptiness; that arose only from the equations of general relativity which are not used in this theory. The relativistic equations were framed on the basis that the law of conservation of matter is valid, but a theory involving continual creation denies that. Hoyle obtained the same result by modifying the equations of relativity to incorporate his C-field.

It is reasonable to enquire about the significance of the time factors in the steady state metric. In models which have a finite past history the expansion makes it possible to have a cosmic time by which the proper times of all fundamental observers can be related. The progressive change in the average density of matter will serve to identify epoch for the fundamental observers. This cannot happen in the steady state universe; its stationary properties do not permit any distinction of cosmic epoch. However the constant velocity/distance relation does permit all fundamental observers, from observations of the recession of the galaxies, to adjust their clocks to go at the same rate. In this way they can all have the same proper time scale though each has his own zero of time upon it. This seems to be the significance of Whitrow's statement that the steady state models involve the concept of cosmic time. [25] Apart from such a cosmological method of correlating clocks, the steady state theory does require all the constants of nature to be real constants and hence atomic clocks, which depend on the vibrations within atoms, cannot change in rate with local epoch and so will run at the same rate everywhere.

Closely connected with considerations of time is that of horizon. There

23. BONDI, H., Loc. Cit. P. 145 f. As Prof. Bondi has recently pointed out to me, the infinitude of an open space is a matter of topology, whereas the infinite character of a closed space is a matter of differential geometry and these are not necessarily connected. Thus on a basis of pattern repetition, if one knows any mesh unit of Euclidean space, one knows them all. In this sense open does not mean just the same as infinite and may be more significant.
24. Here t and T are both in terms of the observer's proper time.
25. WHITROW, G. J., Natural Philosophy of Time, Loc. Cit. P. 262.

is a practical but indefinite horizon due to the limitations of the instruments we can construct. The exponential character of $R(t)$ implies that the velocity of recession approaches asymptotically to the velocity of light, whereby the Doppler effect so degrades the light from the very distant bodies that they become unobservable. In any instrument that can be practically constructed only a finite number of galaxies is visible. But suppose we could have a telescope of unlimited light grasp and unlimited coverage of frequency or wavelength; would there still be a horizon? As in Milne's model there would be a horizon completely occupied by galaxies, theoretically an infinite number of them.

The steady state model has characteristics of space and time determined by its metric which is that of de Sitter. It follows that at great distances time will appear to go more slowly and the horizon is where the observer with an ideal telescope would see time come to a halt, in other words, where a finite period of time at the horizon is seen as infinitely long by the observer. Consequently we have the situation that with an ideal telescope a galaxy once visible would remain visible forever. (P. 130)

In an infinite universe there are fundamental observers whose horizons do not overlap. Between them no influence of any kind can ever pass. Light emitted from one can never reach the other, nor can gravitational fields be propagated through the horizon: the disconnection is complete. ([26])

There are a number of problems raised by the steady state theories, many of which are concerned with the several infinities which it involves. For people who follow an extrapolative line of thought, extending the locally developed laws of science to things on the largest scale, physical infinities are difficult, if not impossible to accept. This attitude is in part based on experience, as mentioned above, and in part is metaphysical; it certainly cannot be lightly set aside. On the other hand those who take a more deductive approach are less likely to have qualms of this kind. Though the objections to actual infinities and to continual creation can be built into a rather strong negative case, there is as much to be said the other way. It hardly seems less objectionable to have a totally inex-

---

26. As Whitrow points out, Loc. Cit., P. 142, a consequence of this infinity of diconnected worlds is that exactly the structure of our solar system, this earth and ourselves must be duplicated infinitely often. Even this book must have been and will be written infinitely often. Operationalists can take a little intellectual exercise over this. On that system of thought it is clearly illegitimate to ask questions about what goes on in disconnected sub-universes but there we have an infinite set of them which are necessarily identical with ours and so presumably questions about them can be answered apart from such as, how to locate them.

plicable singularity (creation) in the past combined with an infinite future existence for matter. In the absence of conclusive observational evidence either way there is room for both lines of thought. If one goes far enough in any scientific investigation metaphysical questions will always arise. In the high days of mechanistic determinism metaphysical or philosophic thought was considered a vice in scientists. Fortunately for the advancement of science and the progress of learning those days have gone.

As does every other theory, that of the steady-state has its problems but it also has some points of appeal. In many ways it leads to the simplest view of the universe of any theory yet developed: it provides for stability on the large scale with evolution on the medium scale and freedom from singularities. As a theory of matter it seems to go more deeply into the nature of the world than any other theory. We have seen the theories in which space is the basic element, expanding and carrying matter with it to finish as a de Sitter world, virtually all space. We have seen the theory of time where the behaviour of light in time determines everything else; almost we could have called it the theory of light. Now we have seen the theory of matter and perhaps matter is more fundamental than space and time. Space is where matter is and time is when things happen. Without matter the other two seem to be shadowy constructions of human minds but these minds, as you know, exist in material shells. And is the steady state theory the theory of matter successful? In its present form that seems most unlikely. Perhaps because of its philosophical implications, it has aroused more controversy than most of the theories and partisans for and against it have argued their cases rather strongly. One can reasonably expect further developments on both sides, but meanwhile the steady state theory and the evolutionary theory are certainly the main contenders with the others now far behind. Cosmology has not yet arrived, if it ever will, at a stage where the ultimate theory can be produced, but there are criteria, apart from logical consistency, which we can apply, at least in principle, to test theories.

There was a Door to which I found no key
There was a veil past which I could not see.

CHAPTER 15

THEORY AND THE WORLD

As there is no possibility of making observations covering the whole universe either in space or in time, we are limited to surveying a part of the universe, which is very large by local standards but not necessarily large on the cosmic scale. If we ask whether this or that model of the universe is a better representation of it than another, one of the questions raised is whether the region we can survey is a fair sample of the whole. Perhaps we can approach this question by starting from the cosmological principle, even though it contains no statement about what is the smallest scale on which it is to be applied. We have seen that in most cosmological models a horizon appears which limits the region of the universe which can, in theory, be surveyed. Sometimes it is a horizon of time, as in the models of de Sitter and Bondi; where time appears to slow down to a halt and so limits the region where events can be observed, the last event taking all eternity to happen. The other kind of horizon is that beyond which the galaxies vanish or into which they merge, beyond which we cannot see. ([1]) Either kind of horizon, like the end of the rainbow, belongs to a particular fundamental observer and must, except for the case of an empty horizon like Eddington's, overlap the horizons of other fundamental observers. From this it does appear that an observer's theoretical horizon contains a region large enough for the cosmological principle to apply.

It is clear that we do not have a prospect of being able to see as far as our theoretical horizon because we cannot construct those ideal telescopes of unlimited power. The question thus becomes one of whether we can see enough of our theoretically observable region to be confident that we are seeing a fair sample. As we are now talking about the observations which we can carry out, it is not legitimate to appeal to spheres of observation overlapping those of other observers at large distances.

In terms of evolutionary models with a finite past history, it was at one time thought that we could see far enough back in time to get close to the beginning of things. When the age of the universe was taken to be

---

1. WHITROW, G. J., *Nat. Phil. of Time*, Loc. Cit. P. 263 ff.

about 1.8 x 10^9 of our years, it was considered that it would call only for a reasonable extension of our observing facilities to reach so far back. However the revised distance scale extended the time scale also and we have no immediate prospect of seeing as far as 1.3 x 10^{10} years. Perhaps at the limit of what is now technically possible, we are reaching towards half of that span. But work at extreme range involves very faint images and so many factors make their interpretations dubious, that our reliable range must be taken as much less than half the age of the universe. Of course this particular question does not arise in the steady state models. Generally, for lack of any evidence to the contrary one does assume that we can see a fair sample, bearing always in mind that it is no more than an assumption.

It may be that we are approaching the stage where we can make observational tests to decide between the predictions of the evolutionary and steady state theories. There are fairly wide spread claims that the latest observations do go against the steady state theory, but it is too early yet to be at sure of this. On two occasions already claims for conclusive disproof of the steady state have evaporated in the light of further investigation. ([2])

In principle we ought to be able to tell whether space is finite or infinite. If it is finite the steady state theory is clearly out, but if space is infinite most of the evolutionary theories are out. There are two ways to go about this. In the flat 3-space of the steady state theory the observed angular diameter of an extended object like a galaxy will decrease with distance in the same way as does the size of an object seen here on earth; the straight line propagation of light provides for that. So the apparent diameters of galaxies at various distances will depend directly on the distance for as far as we can observe. ([3]) If 3-space is spherically curved, the apparent diameter of an object will decrease with distance up to the stage where the curvature begins to have detectable consequences and then its apparent diameter will increase again. The apparent diameter is smallest at $r = 2\pi R$ and thereafter increases because further out space is becoming smaller. If we can see far enough to find the angular diameters of galaxies increasing again we are able to see beyond $2\pi R$ and thus be certain that space is closed. Unfortunately these observations are not as

---

2. BONDI, H., Loc. Cit. P. 43 f.
3. That is out to the horizon. The apparent diameter of a galaxy cannot vanish at a finite distance, though like that of a star within our own galaxy, it can become extremely small.

simple to carry out as they are to state. The images we can obtain are very faint indeed and it is no easy matter to be sure of the Doppler shifts in them which is a very important factor in the measures of luminosity distance. There is no clear result of this test as yet.

The other approach is to examine the distribution in space of galaxies at great distances. The steady state theory predicts a uniform distribution in proper volume, the same at all distances, but in evolutionary theories the galaxies must be seen closer together at great distances than nearby because we are seeing them at an earlier stage of the expansion. This test takes the form of counting, on photographic plates or by means of radio observations, the numbers of galaxies per square degree of a certain apparent brightness and hence of much the same luminosity distance, for several ranges of distance. (^{4}) Of course there are correction factors to be brought in and obviously the greatest luminosity distances are the most significant. Again and most unfortunately, these are the distances where the images are faintest and the results least reliable. In these directions there do not seem to be immediate possibilities of getting very much better results with optical telescopes, but the rapid development of radio astronomy does give promise of greater reliable range in the reasonably near future. Radio telescopes may well prove capable of reaching out further thans $2\pi R$ in terms of closed space and thus be able to make useful measures of angular diameters for more distant objects than we can presently observe. Both for this method and for number counting it is necessary to be able to establish the distances of radio sources. In some cases they can be identified with optically visible objects but it is necessary to go far beyond that and it is by no means certain that light and radio emissions will bear the same ratio at all distances.

There are some other theoretical possibilities for testing rival theories, most of which involve a much greater understanding than we have now of the origin and evolution of galaxies. If we knew as much about the life histories of galaxies as we do of stars, the task would be comparatively simple. (^{5}) It would also be very helpful to know more about the nature and density of matter in the vast spaces between the clusters of galaxies. Here again radio astronomy offers the best hopes for the future.

---

4. There are many thousands of images to be found on such a photographic plate and they have to be evaluated one by one with the aid of an instrument called a microdensitometer, which measures the degree of blackening of the emulsion on the plate for each image. Even with the best instrumental aids, the technique calls for both skill and monumental patience.

5. BONDI, H., Loc. Cit. P. 168 f.

Of all current theories the steady state theory offers itself most freely to refutation. It makes a quite definite prediction that the number of galaxies will be the same per unit proper volume everywhere and at all times. In this respect the evolutionary theories are in a much more sheltered position as they can be modified within rather wide limits to adapt them to observation. Indeed, should observation decide in favour of an evolutionary model, still other criteria will be required to decide which is the most appropriate.

In the last few years a variety of previously unknown objects have been discovered, thanks to rapid advances in astronomical apparatus. It now appears that in our neighbourhood, that is well within our galaxy, there exist stars, or star-like objects, whose strongest observable radiation is in the infra-red region and are thus difficult to examine effectively with conventional telescopes and photographic plates. These infra-red stars, as they are called, may in many cases be stars of known type whose light is very strongly reddened by absorption due to dust clouds lying between us and them. Some stars of this kind have been known for about thirty years, but very recent investigations with powerful infra-red detectors seem to show that there are some objects of very low temperature, below 1000 °K which radiate most of their energy in the infra-red region.

There are other objects which emit strongly in the X-ray range. These have not been detected by ordinary telescopes, partly because X-radiation is very strongly absorbed by our atmosphere and partly because it is not possible to focus X-rays like ordinary light. These new objects have been found by sending detecting apparatus very high into the atmosphere where the absorption is small and determining their positions by scanning with the detectors known areas of sky at known rates. This method, of course, does not yield anything resembling a photographic image. Attempts are now being made by positional calculations to identify these with visible objects. Only two or three out of quite a number have so far been positively identified.

Another approach has been to consider hypothetical "neutron stars", much denser than white dwarf stars, which might generate these rays, but which would barely be visible to the naked eye by their light emission at a few times the distance of the sun. Such objects might conceivably arise from extreme gravitational collapse which, according to theory, may go very much further than one would suppose. The atomic theory has made us accustomed to the idea of matter which even in the densest forms we can handle is still largely empty space. The degenerate matter of white dwarf stars, in which the nuclear particles and electrons are pressed together with no longer any atomic or even nuclear structure, would seem to be the ultimate in contraction, but developments in relativity lead to the strange concept that matter can in certain circumstances collapse much further, even to the extent of virtually going out of existence. ([6]) Meanwhile there seems to be no observational evidence for the physical existence of neutron stars and we can take them to be no more than figments of the theoretician's imaginaton and there are other explanations, such as past supernova explosions for the observed X-rays.

But now cf. Pulsars.

Much more important and intriguing than those are the quasars. ([7]) These objects were discovered by radio telescopes and they show very large red-shifts. If these red-shifts are genuine Doppler shifts and they appear to be so, the quasars have high velocities of recession. A gravitational red-shift as predicted by general relativity would produce the same observed effect as a Doppler shift but it seems to be out of the question as it would mean a mass similar to that of the sun compressed into a sphere only about 10 km in diameter; they would have to be something like neutron stars very much in our astronomical rather than our cosmic neighbourhood. For a variety of reasons this theory cannot be seriously entertained. There is much better evidence that quasars contain sizeable quantities of matter, from about a million times the mass of the sun upwards, according to the various theories about them.

At the time of writing the cosmic status of quasars has not been settled. An obvious proposition is that their red-shifts are true Doppler effects and are the consequence of the expansion of the universe. In terms of the velocity/distance relation, they would have to be regarded as bright young galaxies at vast distances. Thus one of the best known, 3C273, would be at a distance of about 1.5×10^9 light years and have a velocity of recession of 41,000 km/sec. There is even one of them which would be receding at 0.8 of the velocity of light and at a distance approaching 10^{10} light years. Certain of these quasars, notably 3C273, have been identified with optically visible objects ([8]) most notably by Dr. Maarten Schmidt and they do

---

6. In considering the ultimate stages of gravitational collapse one has to reckon with some exceedingly difficult relativistic ideas worked out by Schwarzschild. It appears that if contraction goes beyond a certain stage it is irreversible and ultimately results in the vanishment of the object, however massive, into a point. An eminently readable treatment of this most singular of singularities and other aspects of gravitational theory has been written by ELLIS, G. F. R., *The Theory of Gravitation*, Science Progress 1966 Vol. 54 No. 215 P. 387 ff.
7. The name is an acronym for quasi-stellar radio source. See GREENSTEIN, J. L., *Quasi-stellar Radio Sources*, Scientific American, December, 1963.
8. An interesting account is given by CLARKE, R. W., *Locating Radio Sources with the Moon*, Scientific American, June, 1966.

present the appearance of compact galaxies of a rather blue colour, which would mean that the red-shift has brought emitted shorter wave-length emissions into the visible range.

It was natural to think that quasars are very young galaxies and as they appear to be at large distances, that here was evidence for the evolutionary theories. If this could clearly be shown to be the case it would indeed be conclusive evidence. Unfortunately there are still some unresolved problems here. It is tempting to place the quasars at the beginning of the series of galactic forms as the youngest members, ahead of the spherical and elliptical galaxies, but then there seems to be rather a large gap as the quasars would, in view of the distances assigned to them, have to be all of a hundred times brighter than the others. This might also, though not necessarily, involve a gap in the ages. What is much more serious is that it has been found impossible to account satisfactorily for the enormous output of energy which the quasars would have to produce in order to explain their observed brightness as the remotest galaxies. Thus the quasar 3C273 would have to radiate energy at the rate of 10^{46}erg/sec. Ingenious theories have been developed, for instance by Hoyle and Fowler, [9] to account for the energy by the ultimate in gravitational collapse. This particular theory implies that a quasar should not be regarded as a galaxy in the usual sense of the term but as a super-star, all of one piece as it were. It is not at all obvious how such a prodigious quantity of matter could come to form a single body. So far at least, there does not seem to be any generally accepted theory to account for the required generation of energy by quasars if they really are very distant galaxies.

There is also another remarkable problem for the young galaxy theory. Both the optical and radio outputs of some quasars have been observed to be variable, not with the vast period in which one could expect a change occurring on a scale of galactic dimensions, but with a period in the order of months. Certainly no influence can be propagated through such a body faster than light, which would suggest that the dimensions of quasars should be in the order of a light-month rather than tens or hundreds of thousands of light-years. [10] However one can readily make a model of a quasar in which there is a sort of nucleus having dimensions such that light can span it in a few weeks, surrounded by a very much larger and more diffuse region. The variability could then be a phenomenon involving the nucleus only. A theory of this kind seems to be supported by recent

9. HOYLE, F. AND FOWLER, W. A., Nature, Vol. 197, 1963, P. 533.
10. HOYLE, F., BURBIDGE, G. R. AND SARGENT, W. L. W., On the Nature of Quasi Stellar Sources Nature, Vol. 209, 1966 P. 751 ff.

observations that the variability appears to be confined to the continuous part of the spectrum while the line spectrum which would be expected to emanate from the diffuse part remains stable. ([11])

An alternative theory is that quasars are comparatively local, perhaps about 30 million light-years or less distant, but still quite definitely outside our galaxy. According to this theory, the quasars did originate in our galaxy from which they were expelled with explosive violence. The nucleus of a spiral galaxy such as ours is a rather dense aggregation of stars, hundreds of millions of them concentrated rather like an enormous globular cluster. In the heart of such a nucleus the conditions are very different from those where our sun is situated. If there could be observers in there, the paradox of Cheseaux and Olbers could be a reality and dangerously so. It has even been suggested that one can visualise a considerable number of stars on the verge of disruption, not needing much extra energy to make them become supernovae. In the intense field of radiation prevailing there, such stars are in a situation where an explosive chain reaction might set in, resulting in an enormous galactic outburst in which quite large masses could be expelled with extremely high velocities. ([12]) It seems that just such an explosion may have occurred in our galaxy some 100 million years ago. ([13])

According to this theory the energy requirements of quasars is lower by a factor of a million and extraordinary sources of energy need not be postulated. The kinetic energy which they acquired in the process of expulsion does not make very excessive demands on the galactic explosion as they would be objects of relatively modest size. However, among the large numbers of galaxies which are to be seen on photographic plates one should expect to see evidence of a number of similar explosions, and while this has indeed been found, something else of importance has not been found. It is clear that as a result of explosions in other galaxies around us one ought to find at least a few quasars with strong blue shifts: they could not all be moving away from us. Accordingly this theory does not seem well supported. ([14])

A curious technical problem may arise here. Astronomers are highly experienced in dealing with red-shifts and know how to recognise the

11. I am indebted for this new information to Dr. A. W. Rodgers of Mt. Stromlo Observatory.
12. HUNTER, J. H., SOFIA, S. AND FLETCHER, E., *Quasars - a Minority Report*, Nature, Vol. 210, 1966, P. 346 ff.
13. HOYLE, F., *Recent Developments in Cosmology*, Loc. Cit. P. 112.
14. FAULKNER ET. AL., *Expected Numbers of Blue Shifts and Red Shifts of Ejected Sources*, Nature Vol. 211 1966 P. 502.

pattern of spectral lines even when shifted out of the ultra violet into the visible range. But they have no experience in large blue shifts, even small ones are scarce. Down in the infra-red there are very few sharp lines due to excited atoms, what is found there is mostly rather broad bands due to the more complex chemical molecules. No one knows what a largely blue shifted spectrum looks like and it could take many more years of hard work to find out than it did for red shifts.

A very recent communication may serve to indicate how dubious the position of quasars may be " . . . we must conclude that the red-shifts have nothing to do with distances." [15] Here, Hoyle and Burbidge seem to have made an overpositive statement as so often happens when the situation is unclear. Let us not take this statement out of its proper context but say that, on the balance at present, no theory of quasars has been well established. It may even be that the objects which we call quasars are of several different kinds and at largely differing distances. At this time the cosmological theory that quasars are indeed very distant galaxies seems the more popular and if it can be substantiated we shall, on the present showing, have to adopt some form of evolutionary model for the universe at large. That would still leave to be solved the problem of their enormous output of energy.

And so it seems that this story of cosmology must end in doubt and uncertainty. We have not learned when or how our universe began nor how it will end nor even have we learned for sure the way of it now. It way well be that we cannot ever learn all these things. Has it then been an act of folly that we should have tried to write down the inconceivable and with a formula to encompass the unknowable? I do not think so. It is in the nature of mankind that we cannot bear not to know and that is what science is for, ever to push back the border of the unknown. There have been times when men thought they knew all that was worth knowing, when the dull future of science would be to fill in the small gaps. Well, we have seen something of one aspect of such a period, when nineteenth century science knew it all and we have seen what came of it. Indeed doubt and uncertainty have come of it but also vast stores of knowledge and beautiful concepts and, very importantly, a lesson for science itself. The lesson is that if we have but the wit to see it, knowledge always shows the way to new knowledge. There is a subtle wisdom in the saying attributed, perhaps apocryphally, to Confucius: "It is better to travel hopefully

15. HOYLE, F. AND BURBIDGE, G. R., Nature, Vol. 210, 1966, P. 1346.

than to arrive," for in truth we always deceive ourselves when we think we have arrived in science.

In every field of science there is doubt about the answers to the big questions while certainty and triviality often go together. In this respect there is a marked difference between science and mathematics. In mathematics it is usually possible to ask questions to which there are correct final answers. The questions are based on axioms which have been clearly defined, the rules according to which the game is to be played, and so it is comparatively easy to know whether or not the question is legitimate and, given that it is, we should be able to find the answer. Indeed the answer is contained in a proper question. ([16]) Though science makes use of the mathematical method, the axioms of science have not the same status as those of mathematics, for they are our attempts to set down the rules according to which we think nature works. In terms of these axioms, what are called the "laws of nature," the questions may yield correct answers in terms of them, but there is a court of appeal, nature itself, and as so often happens the answers are rejected because we have not asked the right questions. Thus to make progress in science we must learn how to ask the right questions and this involves continual questioning of our axioms. It is small wonder that there should be doubt and uncertainty and that this is so can be taken as a sign of health and vigour in science.

A little anecdote has been told in which a cosmologist is likened to a man searching for a lost key in the patch of light under a street lamp. Though he did not suppose the key had been lost in that vicinity, he said it was a good place in which to look for a key because of the light there. His attitude was by no means as inconsequential as it may seem and, on the larger scale, it does appear that it is only by light that we can hope to find the key to the universe. In a very real sense, cosmology the grandmother of the sciences can be called the science of light. Cosmology has yet uncountable questions to ask and untold answers to give about the universe. About *the* universe? At the beginning of this book the definition of cosmology was the study of the structure and history of the universe in terms of its large scale features. We can reasonably hope to find keys which will unlock many doors, but shall we find a key to the last door of all, the door of *the* universe? That is too much to hope for but we shall

16. Even in pure mathematics where we might expect to find certainty, Goedel has proved that the logical consistency of a system of mathematics cannot be demonstrated from evidence wholly within that system.

do well to seek the key to *our* universe. That is what Eddington ([17]) meant when he said:

> "We have found a strange footprint on the shores of the unknown. We have devised profound theories, one after another to account for its origin. At last, we have succeeded in reconstructing the creature that made the footprint. And lo! It is our own."

17. EDDINGTON, SIR A. S., *Space, Time and Gravitation*, Cambridge, 1920, P. 131.

APPENDIX I

H. W. M. Olbers on The Paradox, reprinted from

Bode's Jahrbuch, 1826.

Ueber die Durchsichtigkeit des Weltraums, vom Hrn. Dr. Olbers in Bremen, unterm 7. Mai 1823. eingesandt.

Grofs und klein im Raume sind freilich nur relative Begriffe, und wir können uns Geschöpfe gedenken, für die ein Sandkorn so grofs ist, als für uns die ganze Erde; so wie im Gegentheil eine andere Ordnung der Dinge, in der Körper, die die Gröfse ganzer Planeten und Sonnen übertreffen, nur das sind, was uns die kleinsten Sandkörner zu sein scheinen. Aber eben deswegen bleibt es dem Menschen natürlich, die Gröfse oder Kleinheit nach einem Maafsstabe zu beurtheilen, bei dem mittelbarer oder unmittelbarer Weise die Gröfse seines eigenen Körpers, und seiner damit verglichenen nächsten Umgebungen zum Grunde liegt. Nur nach einem solchen Maafsstabe schätzt der Mensch die Gröfse der Dinge, und so mufs er mit staunender Bewundrung die ungeheuern Dimensionen desjenigen Theils des grofsen Weltalls betrachten, der sich nach und nach seinem immer stärker bewaffneten Auge aufschliefst. Schon der Abstand der Sonne von unserer Erde so grofs, dafs man, um sich diese Gröfse begreifli-

Beobachtungen und Nachrichten.   III

cher zu machen, versucht hat, die Zeit zu berechnen, die
eine Kanonenkugel gebrauchen würde, den weiten Raum
zu durchfliegen! Aber dann jeder Fixstern eine Sonne,
und der nächste dieser Fixsterne in einer solchen Ferne
von uns, dafs dagegen der Abstand der Erde von unserer
Sonne fast gänzlich verschwindet! Eine grofse Menge sol-
cher Fixsterne sehr verschiedener Gröfse zeigt sich unse-
rem blofsen Auge, vom blitzenden Sirius bis zu den Sternen 6ter oder 7ter Gröfse, deren Dasein das schärfste Auge bei der heitersten Nacht nur noch kaum ahndet. Viele
dieser kleinen Sterne mögen an sich kleiner sein, als die
gröfser erscheinenden: aber die mehrsten erscheinen uns
doch nur deswegen so yiel kleiner, weil sie so viel weiter entfernt sind, und so sehen wir schon mit blofsem
Auge Sterne, die zwölf- bis funfzehnmal weiter von uns
abstehen, als die Sterne erster Gröfse. Durch Fernröhre
werden uns immer mehr und immer kleinere Fixsterne
sichtbar, je vollkommener diese Werkzeuge sind; und unsere Vernunft mufs zugeben, so schwer es der Einbil-
dungskraft auch fällt, sich so grofse Abstände und Räu-
me noch deutlich vorzustellen, dafs Herschel mit seinen
Riesen Telescopen noch Gegenstände am Himmel erblickte,
die 1500, ja einige tausendmal weiter von uns entfernt
sind, als Sirius oder Arcturus.

Aber ist damit der Scharfblick des nun verewigten
Herrschels den Gränzen des Weltalls nahe, oder auch
nur merklich näher gekommen? Wer kann dies glauben?
Ist der Raum nicht unendlich? Lassen sich Gränzen des-
selben denken? Und ist es denkbar, dafs die schaffende
Allmacht diesen unendlichen Raum leer gelassen haben
sollte? Ich will den grofsen Kant statt meiner sprechen
lassen: „Wo wird die Schöpfung selbst aufhören? sagt
„Kant. Man merkt wohl, dafs um sie in einem Verhält-
„nifs mit der Macht des unendlichen Wesens zu denken,
„sie gar keine Gränzen haben mufs. Man kommt der
„Unendlichkeit der Schöpfungskraft Gottes nicht näher,

112 *Sammlung astronomischer Abhandlungen,*

„wenn man den Raum ihrer Offenbahrung in eine Sphäre
„mit dem Radius der Milchstrafse beschrieben, einschliefst,
„als wenn man ihn in eine Kugel beschränken will, die
„einen Zoll im Durchmesser hat. Alles was endlich ist,
„was seine Schranken und sein bestimmtes Verhältnifs zur
„Einheit hat, ist von dem unendlichen gleich weit ent-
„fernt. Nun wäre es ungereimt, die Gottheit mit einem
„unendlich kleinen Theil ihres schöpferischen Vermögens
„in Wirksamkeit zu versetzen, und ihre unendliche Kraft,
„den Schatz einer wahren Unermäfslichkeit von Naturen
„und Welten unthätig, und in einem ewigen Mangel von
„Ausübung verschlossen zu denken. Ist es nicht vielmehr
„anständig, oder besser zu sagen nothwendig, den Inbe-
„griff der Schöpfung also anzustellen, als er sein mufs, um
„ein Zeugnifs von derjenigen Macht abzugeben, die durch
„keinen Maafsstab kann abgemessen werden? Aus diesem
„Grunde ist das Feld der Offenbarung göttlicher Eigen-
„schaften eben so unendlich, als diese selber sind. Die
„Ewigkeit ist nicht hinlänglich, die Zeugnisse des höch-
„sten Wesens zu fassen, wenn sie nicht mit der Unend-
„lichkeit des Raums verbunden wird."

Soweit Kant. Es bleibt also höchst wahrscheinlich,
dafs nicht blofs der Theil des Raums, den unser auch
noch so stark bewaffnetes Auge übersehen hat, oder übersehen kann, sondern der ganze unendliche Raum mit Sonnen und ihren Gefolgen von Planeten und Kometen be-
setzt ist. Ich sage höchst wahrscheinlich. Gewifsheit kann
uns unsere beschränkte Vernunft nicht geben. Es könnten andere Stellen des Raums ganz andere Schöpfungen
enthalten, als Sonnen, Planeten, Kometen und Lichtstoffe,
Schöpfungen, von denen wir vielleicht gar keinen Begriff
haben können. Halley hat freilich einen Beweis für die
unendliche Menge der Sonnen zu führen gesucht. „Wäre
„ihre Menge nicht unendlich sagt er, so würde sich in
„dem Raum, den sie einnehmen, ein Punkt als der allge-
„meine Schwerpunkt finden, und gegen diesen müfsten
                                                    „sich

Beobachtungen und Nachrichten. 113

„sich alle Weltkörper mit nach und nach beschleunigter „Bewegung stürzen, und also zusammenfallen. Nur weil „der Weltbau unendlich ist. bleibt alles im Gleichgewicht, „und kann sich erhalten." Halley hat blofs an die Schwer-kraft, nicht an Wurfkräfte gedacht. Auch unser Planeten-system würde ja nicht mit der Sonne zusammen fallen, wenn auch gar keine Fixsterne vorhanden wären, wenn es auch ganz isolirt im Weltraum existirte; und dafs Wurf-kräfte unter den Fixsternen wirksam sind, scheint ihre ei-gene Bewegung zu zeigen. Dies wird schon hinreichen, das Unstatthafte des Halleyschen Beweises zu erweisen, gegen den sich auch sonst noch viel erinnern liefse.

Allein, wenn gleich Halley's Beweis nicht gelten kann, so wird es uns doch höchst wahrscheinlich bleiben, dafs die schöne Ordnung, die wir, so weit unsere Seh-kraft irgend reicht, wahrnehmen, auch durch den ganzen unendlichen Raum fortgesetzt sei, und wir haben nur zu untersuchen, ob andere Gründe diese Annahme verwerf-lich machen. Da zeigt sich nun gleich ein sehr wichti-ger Einwurf. Sind wirklich im ganzen unendlichen Raum Sonnen vorhanden, sie mögen nun in ungefähr gleichen Abständen von einander, oder in Milchstrafsen-Systeme vertheilt sein, so wird ihre Menge unendlich, und da müfs-te der ganze Himmel eben so hell sein, wie die Sonne. Denn jede Linie, die ich mir von unserm Auge gezogen denken kann, wird nothwendig auf irgend einen Fixstern treffen, und also müfste uns jeder Punkt am Himmel Fix-sternlicht, also Sonnenlicht zusenden.

Wie sehr dies der Erfahrung widerspricht, braucht wohl nicht gesagt zu werden. Halley läugnet die Fol-gerung, dafs bei einer unendlichen Menge von Fixsternen der ganze Himmel so hell aussehen müsse, wie die Son-ne, aber aus ganz irrigen Gründen. Er verwechselt und verwirrt offenbar scheinbare Gröfse mit der wirklichen, und nur so kann er herausbringen, dafs die Zahl der Fix-sterne zwar wie das Quadrat, ihrer Zwischenräume aber

1826.   H

114 Sammlung astronomischer Abhandlungen,

wie das Biquadrat des Abstandes wachsen. Dies ist nun
ganz irrig. Wenn wir die Fixsterne als gleichförmig im
Weltraume vertheilt voraussetzen, uns mit dem Radius =
1, oder gleich der mittleren Distanz der Sterne erster Gröfse eine Sphäre um unsere Sonne beschrieben vorstellen,
den Halbmesser jedes Fixsterns im Mittel = δ, und ihre
Zahl in diesem Abstande n nennen, so werden sie uns
$\frac{n\delta^2}{4}$ vom Himmelsgewölbe bedecken. In dem Abstande

$= 2$, ist der scheinbare Durchmesser der Fixsterne $= \frac{\delta}{2}$,

aber ihre Zahl $= 4\,n$, und sie werden also wieder von

der Sphäre $\frac{n\delta^2}{4}$ bedecken. So bedecken die in jedem Ab-

stande 1, 2, 3, 4, 5 . . . . m von uns befindlichen Fixsterne gleich viel vom Himmelsgewölbe, und so wird

$$\frac{n\delta^2}{4} + \frac{n\delta^2}{4} + \frac{n\delta^2}{4} \text{ etc.} = m\,\frac{n\delta^2}{4}$$

unendlich grofs werden, wenn m unendlich grofs ist, da

$\frac{\delta^2}{4}$, so klein es auch ist, doch immer eine endliche Gröfse

bleibt. So wird also nicht blofs das ganze Himmelsgewölbe von den Sternen bedeckt, sondern sie müssen noch
hintereinander in unendlichen Reihen stehen, und sich un-
tereinander wieder verdecken. Es ist klar, dafs derselbe
Schlufs statt findet, wenn die Fixsterne nicht gleichförmig im Raume, sondern in einzelne Systeme mit grofsen
Zwischenräumen vertheilt sind.

　　Wohl uns! dafs doch die Natur die Sache anders ein-
gerichtet hat: wohl uns! dafs nicht jeder Punkt des Him-
melgewölbes Sonnenlicht auf die Erde herabsendet. Die
unerträgliche Helligkeit, die alle Vergleichung übersteigen-
de Hitze, die dann herrschen würde, nicht einmal betrach-
tet; (denn für diese, wenn sie gleich über 90,000 mal gröfser sein würden, als wir sie jetzt empfinden, hätte die
schaffende Allmacht unsere Erde und der auf ihr vorhan-

Beobachtungen und Nachrichten. 115

denen Organismen einrichten können) will ich nur der höchst unvollkommenen Astronomie gedenken, die dann uns Erdbewohnern noch möglich bleiben würde. Vom Fixsternhimmel würden wir nichts wissen: unsere eigene Sonne nur mühsam an ihren Flecken entdecken, und blos den Mond und die Planeten als dunklere Scheiben auf dem sonnenhellen Himmelsgrund unterscheiden. Die von dem ganzen, durchaus sonnenhellen Himmel bestrahlten Planeten würden nämlich doch im Verhältnifs ihrer grö-fseren oder kleineren Albedo dunkler erscheinen, als der übrige Himmel.

Aber müssen wir denn deswegen die Unendlichkeit der Fixstern-Systeme verwerfen, weil uns der ganze Himmel nicht sonnenhell erscheint? Müssen wir deswegen diese Fixstern-Systeme nur auf eine kleine Stelle des unendlichen Raums beschränken? — Keinesweges. Bei jener Folgerung aus der unendlichen Menge der Fixsterne haben wir vorausgesetzt, dafs der Weltraum absolut durchsichtig sei, oder dafs Licht, aus parallelen Strahlen bestehend, in jeder Entfernung vom strahlenden Körper ganz ungeschwächt bleibe. Diese absolute Durchsichtigkeit des Weltraumes ist nicht nur ganz unerwiesen, sondern auch ganz unwahr-scheinlich. Wenn gleich die so dichten Planeten durchaus keinen merklichen Widerstand in dem Weltraum leiden, so dürffen wir uns ihn doch nicht ganz leer denken. Manches was wir an Kometen und ihren Schweifen wahr-nehmen, scheint auf etwas materielles im Weltraum hin-zudeuten. Die sich nach und nach zerstreuende Schweif-materie der Kometen und der Stoff des Thierkreislichtes sind doch gewifs darin vorhanden. Selbst wenn dieser Weltraum auch sonst ganz leer wäre, müssen und können die sich durchkreuzenden Lichtstrahlen einen kleinen Ver-lust bewirken. Dies scheint nicht nur a priori erweislich, man mag, nun Newtons, oder Huygens Hypothese über die Natur des Lichts annehmen: sondern es wird auch durch die Vergleichung der Cassegrainschen und Gre-

H 2

116 Sammlung astronomischer Abhandlungen,

gorianischen Telescope, und der relativen Dichtigkeit des Lichts vor und hinter dem Brennpunkt sphärischer Spiegel bestättiget *).

Gewifs ist also der Weltraum nicht ganz absolut durchsichtig. Aber es bedarf nur eines äufserst geringen Grades von Undurchsichtigkeit, um jene, der Erfahrung so ganz widersprechende Folgerung aus einer unendlichen Menge von Fixsternen, dafs dann der ganze Himmel uns Sonnenlicht zurücksenden müsse, völlig zu vernichten. Nehmen wir zum Beispiel an, der Weltraum sei nur in dem Grade durchsichtig, dafs von 800 Strahlen, die Sirius ausstrahlt, 799 bis zu der Entfernung gelangen, worin wir uns von ihm befinden, so wird schon dieser ganz kleine Grad von Undurchsichtigkeit mehr als hinreichend sein, das unendlich ausgedehnte Fixsternsystem uns so erscheinen zu lassen, wie wir es wirklich sehen.

Da aus allen Punkten der Oberfläche leuchtender Körper Lichtstrahlen in jeder Richtung ausströmen, so können wir uns dieses Licht in einzelne, aus unter sich parallelen Strahlen gebildete Strahlen-Cylinder getheilt vorstellen. Die Helligkeit des leuchtenden Körpers wird dem Auge im Verhältnifs der Dichtigkeit des Lichts in diesen Strahlen-Cylindern erscheinen. Nun verhält sich, nach den Gesetzen, wie das Licht bei seinem Fortgange in nicht abso-

*) Philosophical Transactions. Year 1813. 1814. Bei Berechnung der relativen Dichtigkeit des Lichts vor und hinter dem Brennpunkt concaver Spiegel scheint Capit. Kater nicht daran gedacht zu haben, dafs man den sogenannten Brennpunkt hier nicht als einen physischen Punkt betrachten darf, sondern dafs sich dort das Bild der Sonne oder der Kerzenflamme befindet. Dies wird in den Rechnungen einige Correctionen veranlassen, aber das Resultat, dafs Licht beim Durchgange durch den Brennpunkt verloren gehe, nicht aufheben. Es ist sehr zu wünschen, dafs diese interessanten Versuche, die sich vielleicht noch zweckmäfsiger einrichten liefsen, sorgfältig wiederholt werden mögen.

Beobachtungen und Nachrichten. 117

lut durchsichtigen homogenen Substanzen geschwächt wird, bei jedem unendlich kleinen Fortgange die Abnahme der Dichtigkeit des Lichts, wie diese Dichtigkeit selbst. Es sei also die Dichtigkeit des Lichts in dem Abstande x, vom strahlenden Körper $= y$, so wird es, indem es um dx weiter fortrückt, um dy geschwächt, und es ist

$$dy = - \, a y d x$$

oder integrirt

$$\log y = \text{Const.} - ax.$$

Die Constante wird dadurch bestimmt, dafs $y = A$ ist, wenn $x = o$, und so haben wir die Gleichung

$$\log \frac{y}{A} = - ax.$$

Hier ist nun, wenn $\log \frac{y}{A}$ der natürliche Logarithmus bleibt, a gleichsam das Maas der Undurchsichtigkeit des Weltraums, und $\frac{1}{a}$ die Subtangente der logarithmischen Linie, nach deren Ordinaten die Helligkeit des gesehenen Gegenstandes mit der Entfernung abnimmt. Bei Rechnungen über das Verhältnifs von A:y können wir für $\log \frac{y}{A}$ den künstlichen Logarithmen gebrauchen, und müssen uns uur erinnern, dafs alsdann auch a das mit 0,43429448... multiplicirte Maas der Undurchsichtigkeit ist.

Nach obiger freilich willkührlichen Annahme, dafs das Licht eines Fixsterns, wenn er so weit als Sirius von uns absteht, bis auf $\frac{799}{800}$ geschwächt werde, will ich nun a suchen. Es sei also der Abstand des Sirius $= 1$, so ist

$$\log 799 \ldots 2.9025467793$$
$$\log 800 \ldots 2.9030899870$$
$$\overline{ a = 0.0005432077}$$

also ist $\log a = 6{,}7349604 - 10$. Damit läfst sich nun leicht berechnen, wie die Helligkeit der Fixsterne mit ihrer weitern Entfernung von uns abnimmt. Setzt man A,

118 *Sammlung astronomischer Abhandlungen,*

oder die Helligkeit unserer Sonne gleichfalls $= 1$, so ist die Helligkeit eines Fixsterns noch

$\frac{9}{10}$ in dem Abstande von   84,23 Sirius Distanzen
$\frac{8}{10}$ — — — — — 178,40 — — —
$\frac{7}{10}$ — — — — — 285,16 — — —
$\frac{6}{10}$ — — — — — 408,41 — — —
$\frac{5}{10}$ — — — — — 554,13 — — —

Man sieht also, dafs bis auf alle Distanzen, in denen unser bewafnetes Auge noch einzelne Fixsterne unterscheiden kann, die Helligkeit nur bis auf $\frac{1}{2}$ abnimmt. So grofse Unterschiede, und noch gröfsere mögen in der absoluten Helligkeit der Fixsterne selbst statt finden.

Man mufs durchaus nicht Helligkeit mit Lichtstärke verwechseln. Die Lichtstärke ist nemlich die Helligkeit mit der scheinbaren Gröfse multiplicirt, und die Lichtstärke verhält sich also direct wie die Helligkeit, und zugleich verkehrt wie das Quadrat des Abstandes. So hat ein Stern, der 554mal weiter als Sirius von uns entfernt ist, zwar noch die halbe Helligkeit, aber weniger als $\frac{1}{610000}$ der Lichtstärke des Sirius.

In gröfsern Distanzen nimmt nun die Helligkeit sehr ab. In dem Abstande von 1842,9 Sirius Weiten ist sie nur noch $\frac{1}{10}$, in dem Abstande 3681,8 nur $\frac{1}{100}$, in dem Abstande 5522,7 nur $\frac{1}{1000}$ u. s. w. der ursprünglichen Helligkeit.

In welchem Abstande hat ein Fixstern noch die Helligkeit des Vollmondes, diese $= \frac{1}{300000}$ der Helligkeit der Sonne gesetzt? Es ist

$$\log \tfrac{1}{300000} = -\, 5{,}4771213$$
davon ist wieder der Log. $= 0{,}7385524$
$$\underline{\log a = 6{,}7349604 - 10}$$
$$\log x = 4{,}0035920$$

giebt $x = 10083{,}05$. Also in dem Abstande von 10000 Sirius Weiten ist die Helligkeit der Fixsterne nur noch so grofs, als die des Vollmondes. Es werden eine ungemein grofse Menge so entfernter Sterne in einem sehr dichtem

Beobachtungen und Nachrichten.   119

Sternhaufen vereiniget sein müssen, wenn wir einen solchen Sternhaufen bei der heitersten mondlosen Nacht noch als einen blafsen Nebelfleck mit unsern vollkommensten Fernröhren unterscheiden sollen.

Unsere vom Vollmond erleuchtete Atmosphäre hat für uns noch nicht $\frac{1}{750000}$ der Helligkeit des Vollmondes selbst, und ist doch noch grofs genug, dem blofsen Auge alle Sterne, die weniger als die 4te oder 5te Gröfse haben, unsichtbar zu machen.   Um zu sehen, in welchem Abstande die Fixsterne noch so hell sind, als der Grund des Himmels in einer Vollmondsnacht, nehme man

$$\log (300{,}000 \times 90000) = 10{,}4313638$$
$$\text{davon ist der } \log \ldots 1.0183410$$
$$\log a \ldots 67349604 - 10$$
$$\overline{\log x = 4.2833806.}$$

Giebt x = 19203,5.

Ich will nun noch die Helligkeit eines Fixsterns berechnen, der 30000 Sirius Weiten von uns absteht.

$$\log x = 4{,}4771213$$
$$\log a = 6{,}7349604 - 10$$
$$\overline{\log ax = 1.2120817.}$$

Dazu gehört die Zahl 16,29602.... Also ist $\log \frac{y}{A}$ $= -16{,}29602.$ ...... Dies ist der Log. von 1977100000 Millionen, und um so viel mal ist die absolute Helligkeit des Fixsterns geschwächt. Um dies ungeheure Verhältnifs etwas fasslicher zu machen, kann man bemerken, dafs diese dem Fixstern noch bleibende Helligkeit 65900 Millionen mal schwächer ist, als die Helligkeit des Vollmonds, oder 732250 mal schwächer, als die Helligkeit des Himmelgrundes in einer heitern Vollmondsnacht, welches man allerdings schon als völlig dunkel ansehen kann.

Wir können also sicher annehmen, dafs unter diesem vorausgesetzten Grade der Durchsichtigkeit des Weltraumes, alle Sterne, die über 30000 Sirius-Weiten von uns

120 *Sammlung astronomischer Abhandlungen,*

abstehen, nichts mehr zur Helligkeit des Himmelgrundes
beitragen.

Dieser Grund des Himmels würde uns also völlig dun-
kelschwarz erscheinen, wenn nicht unsere eigene Atmosphäre auch blofs von den Sternen erleuchtet, schon einige
Helltgkeit hätte, die auch in der heitersten Nacht den Grund
des Himmels nicht völlig schwarz, nur dunkelblau erscheinen läfst.

Dafs der Grund des Himmels wirklich ganz schwarz
aussehen, wirklich ganz ohne alles merkliche Licht sein
würde, wenn wir nicht durch unsere, vom Sternenlicht
erleuchtete Atmosphäre sehen müfsten, scheint schon eini-
germafsen aus dem zu folgen, was wir an der Venus wahr-
nehmen. Der von der Sonne nicht erleuchtete Theil ihrer Scheibe wird nur zuweilen durch ein eigenes, phos-
phorisches Licht, also dadurch erkennbar, dafs er heller
ist, als der übrige Himmelsgrund; nie dadurch, dafs er
dunkeler ist, als der übrige Himmelsgrund, von dem er
doch einen Theil bedeckt. Der bedeckte Theil dieses Himmelgrundes ist also merkbar um nichts dunkler, als der
unbedeckte. Dasselbe läfst sich auch beim Mars wahrnehmen, wenn dieser nicht ganz erleuchtet ist. — So beschrei-
ben auch diejenigen, die auf hohen Bergen den gestirnten
Himmel zu betrachten, Gelegenheit hatten, den Grund des
Himmels schon als sehr dunkel, ja völlig schwarz, obgleich
sie noch durch den gröfseren Theil unserer Erdatmosphäre
hindurch sehen mufsten.

Ich weifs nicht, ob ich mich täusche, aber es ist mir
oft vorgekommen, als wenn unter kleinen Fixsternen von
einerlei Lichtstärke (die Lichtstärke ist nemlich, wie schon
oben erinnert, die Helligkeit mit der scheinbaren Gröfse
multiplicirt) einige ein blitzendes, scintillirendes, andere
ein stilles, ruhiges Licht hätten. Wenn dies keine Täu-
schung ist, so würde ich geneigt sein, die ersten für näher und kleiner, die anderen für an sich und scheinbar
gröfser, aber entfernter zu halten, davon durch die Un-

Beobachtungen und Nachrichten. 121

durchsichtigkeit des Weltraums geschwächtes Licht nicht mehr die zum Scintilliren erforderliche Dichtigkeit hat.

Die Annahme, dafs das Licht unabhängig von seiner Divergenz, indem es vom Sirius bis zu uns kommt, um $\frac{1}{800}$ geschwächt werde, ist natürlich ganz willkührlich. Ich habe, wie gesagt, blos dabei beabsichtigt, zu zeigen, dafs schon ein so geringer, ja ein noch geringerer Lichtverlust auf diese grofse Distanz hinreichend sei, die Erscheinungen am Himmel so darzustellen, wie wir sie sehen, wenn auch die Menge der Sterne durch den ganzen unendlichen Raum unendlich ist. Ganz ohne alle Ueberlegung ist indessen dieser Grad von Undurchsichtigkeit für den Weltraum nicht gewählt, und ich glaube, dafs er von dem wirklich statt findenden nicht so ganz aufserordentlich verschieden sein dürfte.

So hat also mit weiser Güte die schaffende Allmacht den Weltraum zwar in einem ungemein hohen Grade, aber doch nicht absolut durchsichtig gemacht, und so unsere Sehkraft auf einen bestimmten Raum des unendlichen beschränkt: da wir nur dadurch in den Stand gesetzt sind, etwas von dem Bau und der Einrichtung des Weltalls kennen zu lernen, von dem wir wenig wissen würden, wenn auch die entferntesten Sonnen ganz ungeschwächtes Licht zu uns schicken könnten.

APPENDIX II

J. P. L. de Cheseaux on The Paradox, reprinted

from his "Traité de la Comète", Lausanne 1744

Sur la force de la Lumiére & sa propagation dans l'Ether , & sur la distance des Etoiles fixes.

C'est une Propofition démontrée dans l'Optique, que fi toutes les Etoiles fixes é-toient autant de Soleils égaux & femblables au nôtre, en forte que placées à la même diftance, elles paruffent fous un même dia-mètre, & avec un éclat de lumiére égal au fien, ou qu'elles nous envoyaffent la même quantité de lumiére, il eft, dis-je, démontré, que la quantité de lumiére que chacune d'elles placée à quelque diftance de la terre que ce foit, nous envoyeroit, feroit à celle du Soleil, en raifon directe du quarré de fon diamètre, apparent au quar-ré de celui du Soleil, ou en raifon inver-fe du quarré de fa diftance au quarré de celle du Soleil. Concevant maintenant tout l'efpace étoilé, divifé en couches fphériques, concentriques, & d'une épaiffeur à peu près conftante, égale à celle du tourbillon ou Syf-tème Planétaire de chaque Etoile ; & fup-pofant le nombre des Etoiles contenu dans chaque couche, à peu près proportionnel

à

224 *T R A I T E'*

à la furface de cette couche, ou, au quarré de fa diftance au Soleil, pris pour centre de tout l'efpace étoilé; & enfin, les
diamètres véritables de chaque Etoile à peu-
près égaux à celui du Soleil, comme je
l'ai fuppofé dès le commencement, on trou-
vera la quantité de lumiére qui nous eft envoyée par les Etoiles de chaque couche proportionelle à la fomme des quarrés de leurs
diamètres apparens, c'eft-à-dire, proportionelle au nombre des Etoiles de chaque couche , multiplié par le quarré du diamètre
apparent de chacune, ou par ce que je viens
de dire, proportionelle au quarré de la
diftance de chaque couche divifé par ce
même quarré ; & par conféquent, cette
quantité de lumiére toûjours la même pour
toutes les couches ; & chacune aura à la
quantité de lumiére que nous recevons du
Soleil, le rapport conftant du quarré de
la diftance du Soleil à la Terre, au quarré
de la diftance de la prémiére couche divifé par le nombre des Etoiles contenues dans
cette couche, c'eft-à-dire, à peu près le
rapport de I *α* 4 000 000 000. De là
il fuit que fi l'efpace étoilé eft infini, ou
feulement plus grand que la prémiére couche, y compris le tourbillon du Soleil dans

la

DE LA COMETE. 225

la raiſon du cube de 760 000 000 000 000. à 1., chaque point du Ciel nous pa-roitroit auſſi lumineux qu'un point du So-leil de même grandeur apparente, & par conſéquent la Lumiere que nous recevrions de celui des deux Hémiſphères Céleſtes qui eſt ſur notre Horiſon ſeroit 9 1 8 5 0. ſois plus grande que celle que nous recevons du Soleil. La différence énorme, qui ſe trouve entre cette concluſion & l'expérience, fait voir, ou que la Sphère des Étoiles fixes, non ſeulement n'eſt pas infinie, mais même qu'elle eſt incomparablement moindre que l'étenduë finie que je lui ai ſuppoſée, ou que la force de la lumiére décroit en plus grande proportion, que la raiſon inverſe des quarrés des diſtances. Cette derniére ſuppoſi-tion eſt aſſez vrai-ſemblable, elle demande ſeulement que l'eſpace étoilé ſoit rempli de quelque fluide, capable d'intercepter, tant ſoit peu, la lumiére. Quand ce flui-de ſeroit 330 000 000 000 000 000. plus tranſparent ou plus rare que l'eau, il ſuffiroit pour affoiblir la force de la Lu-miére d'une 33me. partie à ſon paſſage par chaque couche, & pour abſorber même par degrés toute la Lumiére, qui nous eſt envoyée par toutes les Etoiles qui ſont au de-

Tome I. P là

226   *T R A I T E'*

là des couches voifines de notre Tourbillon, au poiut de réduire celle de tout un
Hémifphère à la 430 000 000me. partie
de la quantité de Lumiére que nous recevons du Soleil, ou à une quantité de Lumiére feulement 33 fois plus grande que
celle que nous recevons du globe obfcur
de la nouvelle Lune éclairée par la Terre.

ON jugera bien, fans doute, que les
nombres que j'ai pofés, font pris par con-
jeéture : Cela eft vrai, mais ces conjeétu-
res ne font pas arbitraires, au moins la prémiére fur la diftance des Etoiles fixes de la
prémiére grandeur, que j'ai pofée environ
240 000. fois plus grande que celle du
Soleil; & cela, fur ces principes.   J'ai
remarqué que la fplendeur de Mars au tems
de fa Conjonétion avec Saturne, le 16, 17,
18. May 1743, furpaffoit celle de toutes les
Fixes de la 1ere. grandeur, même de Sirius;
ce qui, en ayant égard à fa diftance à la
Terre, ou à fon diamètre apparent, à fa
figure dichotome, à fa diftance au Soleil,
ou à l'intenfité de fa Lumiére, comparée à
l'intenfité de celle de la Lune & du Soleil,
m'a donné le diamètre apparent des Etoiles
de la prémiére grandeur, moindre que la
$\frac{1}{119}$ partie d'une feconde de degré.   Au

DE LA COMETE.   227

tems de fa Conjonction avec Jupiter, le
1er. & le 2 de Juin, fa fplendeur paroif-
foit égale à celle de Regulus, ou un peu
moindre que celle des plus grandes Etoiles : ce qui m'a donné, en fuivant les mêmes précautions que la prémiére fois, le
diamètre des Fixes de la 1ere. grandeur un

peu plus grand que la $\dfrac{1}{131}$ partie d'une

feconde.   Donc par un milieu, le diamè-

tre de ces Etoiles fera la $\dfrac{1}{125}$ partie d'une

feconde, ou de 0° 0′ 0″ 0‴ 28^{IV} 48″
Ce diamètre comparé avec celui du Soleil,
la parallaxe du Soleil prife de 1 5″, l'o-
pacité de l'eau déterminée par Mr. B o u-
G U E R dans fon Effai fur la gradation de
la Lumiére, m'ont donné en fuivant les
méthodes de ce favant Académicien, pref-
que fans autre fecours, tous mes nombres
à l'exception de la fuppofition fuivante,
que j'ai été obligé de faire fur la proportion de la quantité de Lumiére envoyée
par les feules Etoiles de la prémiére couche, à celle qui eft envoyée par tout le
Firmament.   Le diamètre de Jupiter Pe-
rigée, étant de 50″, & fa diftance au Soleil 5 fois plus grande que celle de la
P   2   Lune.

228   *T R A I T E'*

Lune , & par conféquent l'intenfité de fa
Lumiére 7 5 0 0 0 0 0 fois plus foible que
celle du Soleil ou des Etoiles fixes : j'ai
trouvé que , pofant le diamètre apparent
de celles de la prémiére grandeur de $0°$
$0'$ $0''$ $0'''$ $28''''$ 48^{V}, la fplendeur de Jupiter égaloit (dans ces circonftances) celle de 5 de ces Etoiles , ou , qu'elle étoit
égale au $\dfrac{2}{5}$ ou au $\dfrac{1}{3}$ de toutes celles du
prémier rang, prifes enfemble. Or la fplen-
deur de cette Planete , m'a paru , autant
que j'en ai pu juger par quelques expé-
riences grofliéres, en comparant certains
objets éclairés par un certain efpace du
Firmament , avec ces mêmes objets éclairés
par la lumiére de cette Planete feule; la fplendeur de Jupiter, m'a paru , dis-je , environ
la 50me. partie de celle de tout un Hémifphère , ou la 100me. de tout le Firmament.   D'où j'ai conclu cette derniére
3 3 fois plus grande que celle des Etoiles
de la 1ere. grandeur ou de la prémiére couche.   On peut remarquer que celle de Venus Dichotome , & dans fes moyennes dif-
tances, fe trouve environ 4 fois plus grande que celle de Jupiter , & par conféquent,

                                        fuivant

D E   L A   C O M E T E.   229

fuivant ces hypothèfes, la douziéme partie de celle de tout un Hémifphère.

I I I.

NAME INDEX

SUBJECT INDEX